The Dots-and-Boxes Game

The Dots-and-Boxes Game
Sophisticated Child's Play

Elwyn Berlekamp

To Alvin,
Elwyn Berlekamp
March 22, 2014

A K Peters
Natick, Massachusetts

Editorial, Sales, and Customer Service Office

A K Peters, Ltd.
5 Commonwealth Road, Suite 2C
Natick, MA 01760

Copyright © 2000 by A K Peters, Ltd.

All rights reserved. No part of the material protected by this copyright notice may be reproduced or utilized in any form, electronic or mechanical, including photocopying, recording, or by any information storage and retrieval system, without written permission from the copyright owner.

Library of Congress Cataloging-in-Publication Data

Berlekamp, Elwyn, R.
    The dots-and-boxes game : sophisticated child's play / Elwyn Berlekamp
    p.cm.
    ISBN 1-56881-129-2 (pbk. : alk. paper)
    1. Game theory. 2. Dots-and-boxes (Game) I. Title.

QA269 .B39 2000
519.3–dc21   00-033185

Printed in the United States of America
15 14 13 12 11   10 9 8 7 6

Dedicated to
Persis, Bronwen, and David

Table of Contents

Preface   ix

1. Dots-and-Boxes—An Introduction   3
2. Strings-and-Coins   11
3. Elementary Chain Counting Problems   13
4. Advanced Chain Counting   23
5. Advanced Chain Counting Problems   27
6. Nimber Values for Nimstring Graphs   39
7. Elementary Problems with Nimbers   53
8. More about Nimstring, Arrays, Mutations, Vines, etc.   59
9. Advanced Nimstring Problems   69
10. Playing Dots-and-Boxes with Very Close Scores   75
11. Dots-and-Boxes Problems with Close Scores   87
12. Unsolved Problems   117

Bibliography   125

Index   127

Preface

Like many other children, I learned to play the game of Dots-and-Boxes soon after I entered grade school. That was in 1946. Ever since then I have enjoyed recurrent spurts of fascination with this game. During several of these bursts of interest, my playing proficiency broke through to a new and higher plateau. This phenomenon seems to be common among humans trying to master any of a wide variety of skills.

In Dots-and-Boxes, however, each advance can be associated with a new mathematical insight! Players on each plateau all share some key insight which remains unknown to those on lower levels.

New results continue to be discovered every few years. One of my graduate students and I found another one less than a month ago. Yet more discoveries surely remain ahead.

Although the theorems are mathematical, many of them can be expressed in terms that can be understood and used by grade school children, and most of the proofs can be mastered by a good high school math student. Yet the search for these fascinatingly simple results has challenged some of the world's foremost mathematicians for several decades. I am confident that most readers of this book will soon share my view that Dots-and-Boxes is the mathematically richest popular child's game in the world, by a substantial margin.

When I was a child, I played as a child. Not until junior high school did I discover the power of double-dealing moves (Chapter 1). Although my Dots-and-Boxes play remained at that same level for the next ten years, I did become an avid fan of Martin Gardner's "Mathematical Games" column in *Scientific American*. I was awestruck when someone first showed me Bouton's solution to the game of Nim, and awestruck again when I first read some of the work of Richard Guy and Cedric Smith. The fact that some games have mathematically based winning strategies made a big impression on me, although I did not yet perceive any hope of finding such structure in Dots-and-Boxes.

When I was a student at MIT, I wrote bridge columns for *The Tech* and became known as a games buff. One day some other students whom I did not know recruited me to play a demonstration game against a rudimentary computer program they had written to play 3 × 3 Dots-and-Boxes. That program had superb graphics for its time, although the algorithms were primitive and the processor was slow. Nevertheless, the program beat me. This certainly captured my attention! Within the next few days, I discovered the importance of the parity of the number of long chains. This moved me on to the next plateau, but instead of staying there, I continued to focus on the 3 × 3 game. I discovered the "swastika" strategy explained on page 9 of this book. Less than a week after my initial defeat, I returned to resume the competition with the computer. I won all of the next ten games: five going first, and five going second.

Conventional "computer game-playing" approaches do not work very well on Dots-and-Boxes. Progress early in the game is very hard to measure because, as explained in Chapter 1, the most important issue of the fight is whether the number of long chains will be odd or even. On big boards, neither player knows whether he should try to increase or decrease this number. Often that remains unclear, even after the game reaches the stage at which it is feasible to obtain a complete solution to a closely related game described in Chapter 6. In recent years, computers have become so fast that some programs based on traditional artificial intelligence techniques can now play at a respectable level on smaller boards. One of the best, written by J. P. Grossman, can be found at http://dabble.ai.mit.edu/

While studying electrical engineering at MIT, I learned about duality of planar graphs, and observed that Dots-and-Boxes could also be described in terms of a completely equivalent dual game, Strings-and-Coins. Yet, throughout my years in graduate school, as an assistant professor at Berkeley, and visiting academic stints at UNC and USC, my sporadic bursts of Dots-and-Boxes play remained on the same plateau. I could then have solved some, but not all, of the problems in Chapter 5 of this book.

Then I met Richard Guy and began hunting for ways to extend his beautiful work on Sprague-Grundy theory. This led me to a new game, which I christened "Nimstring"! I composed a number of problems including "vines," especially "Kayles-vines," and presented them at a talk Richard Guy hosted at the University of Calgary. As an applied mathematician, I was very proud to be able to relate the beautiful but "pure" work of Bouton, Sprague, Grundy, Smith, and Guy to a very popular child's game which had long seemed so unsophisticated. In the late 1960s, I met another avid student of mathematical games named John Horton Conway. We spent several days discovering that there were several games in which each of us could nearly always beat the other.

We agreed to join forces with Richard Guy to write a book, and began in the late 1960s with hopes of completing it in two or three years. But Conway soon discovered the theory of partizan games. The goals of the project expanded and expanded, until the comprehensive, two-volume first edition of *Winning Ways* [1] was published in 1982, with Chapter 16 devoted to the game of Dots-and-Boxes. Some of the results, including most of our "Harmless Mutation Theorem," were primarily due to Richard and John. As a computer scientist at that time, I tried hard to find as many "NP" game results as I could. The only two I found which dealt with games both appeared in *Winning Ways*. One of them deals with loony endgames at Dots-and-Boxes. A refined version that is more applicable to real play appears in Chapter 11 of this book.

About every two or three years since 1971, I have taught courses or seminars on mathematical games at UC Berkeley. Many topics that began as student projects in this course evolved into significant contributions to the subject. Many of these works are described in references [1], [5], [7], and [8]. I usually spend two or three weeks on Dots-and-Boxes. While teaching this course, I found a considerable gap between Nimstring theory (especially the composed "vine" problems which illustrate the power of this theory so well) and the advanced chain-counting techniques of Chapter 4, which prove so powerful in actual play. The first published reconciliation of these two viewpoints appears on pages 50–52 of this book. It is a powerful technique which will allow you to compute nimstring values of realistic positions more quickly.

In addition to my enormous debt to Richard Guy and John Conway, I must mention several others, without whose help over the years this book would not exist. Timothy Riggle was my favorite Dots-and-Boxes opponent in grade school. We moved to different cities when we were 9, and lost contact for the next 40 years. We then ran into each other again at a math conference in 1989. Tim was then chairman of math and computer science at Baldwin Wallace College, and he then got them to invite me to give the Regents' Lecture there in 1992. In preparation for my visit, Tony Lauria organized a math games tournament in which over 100 students participated. The success of this tournament inspired me to do likewise. Since then, I have sponsored tournaments on particular games at several math conferences, including one whose final games were published [7]. These tournaments have led to an increased popularity of the

5 × 5 board. It is big enough to be quite challenging, and yet small enough to keep games reasonably short.

John L. Kelly, Jr., who was my boss during two of my first three summers at Bell Labs, showed me Bouton's elegant solution to the game of Nim. Most recently, several current Berkeley graduate students have contributed to studies of Dots-and-Boxes. Freddy Mang wrote a very helpful computer program described further at the end of Chapter 6. Saul Schleimer and Katherine Scott have become so proficient at 5 × 5 Dots-and-Boxes that from time to time they beat me. This book began as a collection of interesting positions that occurred in those games. Apollo Hogan began an early on-line version of this collection.

The present form of this book benefited greatly from the encouragement of Alice and Klaus Peters, and especially from the care and attention of their production editor, Ariel Jaffee. My initial title was "100 Dots-and-Boxes Problems with Solutions". They persuaded me to append the appropriate text. Some topics, which had already been explained so eloquently by my co-authors in *Winning Ways*, were shamelessly copied. Other topics from the Dots-and-Boxes chapter of *Winning Ways* were omitted, replaced, or expanded.

In some circles, Dots-and-Boxes remains as popular today as ever. I've encountered strangers playing games between children and children, between parents and children, and between two parents, at places ranging from Germany to airplanes to Parents' Day events at high schools in Connecticut, Kentucky, and California. Nearly all of these games are played on only the most primitive level of skill.

In fact, perfect play at 3 × 3 Dots-and-Boxes is simpler than perfect play at 3 × 3 Tic-Tac-Toe. Yet the latter is known by many; the former, by very, very few. Chapter 1 of this book aims to bridge that gap.

Mathematics is all around us. It lurks just beneath the surface of many things. Even though many of the results are relatively easy to explain, they have long gone mostly unnoticed. This book reveals some of the mathematical notions that will enable you to become an expert at Dots-and-Boxes.

Conventions for Problems

A (Arthur) = first player; *B* (Beth) = second player. Arthur plays the odd-numbered turns; Beth, the even.

The small number shown below the lower left (southwest) corner of each problem is the count of the number of moves already played. Although boxes have already been made in some problems, *there have been no doublecrosses yet.* We use the geographic terms to describe different regions of the board (e.g. Northwest is the top left).

We also use "chain" to mean "long chain" unless otherwise specified.

Chapter 1
Dots-and-Boxes—
An Introduction

Dots-and-Boxes is a familiar paper and pencil game for two players; it has other names in various parts of the world. Two players start from a rectangular array of dots and take turns to join two horizontally or vertically adjacent dots. If a player completes the fourth side of a square (box) he initials that box and must then draw another line. When all the boxes have been completed the game ends and whoever has initialed more boxes is declared the winner.

A player who can complete a box is not obligated to do so if he has something else he prefers to do. Play would become significantly different were this obligation imposed.

Figure 1 shows Arthur's and Beth's game, in which Arthur started. Then, Beth's big sister, Amy, played the game shown in Figure 2 against Arthur's friend Babar. This time Amy played first. Nothing was given away in the fairly typical opening that Arthur and Beth had used. Babar was happy to copy Beth's replies from that game and was delighted to see Amy follow it even as far as that unlucky thirteenth move, which had proved Arthur's undoing. Babar grabbed those 2 boxes and happily surrendered the bottom 3, expecting 4 in return. But Amy astounded him by giving him back 2. He pounced on those, but when he came to make his bonus move, realized he was double-crossed!

The Dots-and-Boxes Game

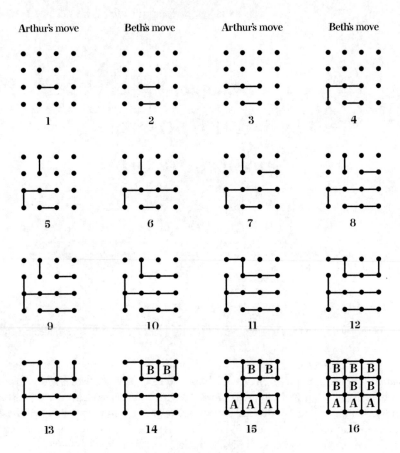

Figure 1. Arthur's and Beth's game.

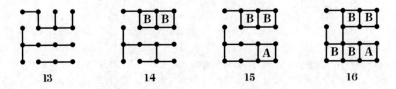

Figure 2. Amy's brilliance astounds Babar.

Amy beats all her friends in this double-dealing way. Most children play at random unless they've looked quite hard and found that every move opens up some chain of boxes. Then they give the shortest chain away and get back the next shortest in return, and so on.

But when you open a long chain for Amy, she may close it off with a double-dealing move, which gives you the last 2 boxes but forces you to open the next chain for her (Figure 3). In this way she keeps control right to the end of the game.

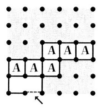

Figure 3. Amy's double-dealing move.

You can see in Figure 4 just how effective this strategy can be. By politely rejecting two cakes on every plate but the last you offer her, Amy helps herself to a resounding 19 to 6 victory. In the same position you'd have defeated the ordinary child 14 to 11.

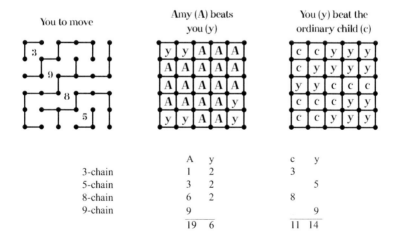

	A	y		c	y
3-chain	1	2		3	
5-chain	3	2			5
8-chain	6	2		8	
9-chain	9				9
	19	6		11	14

Figure 4. Double-dealing pays off!

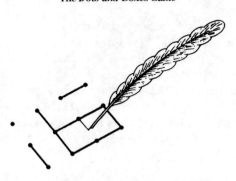

Figure 5. A doublecross—two boxes at a single stroke.

Double-Dealing Leads to Double-Crosses

Each double-dealing move is followed, usually immediately, by a move in which two boxes are completed with a single stroke of the pen (Figure 5). These moves are very important in the theory. We'll call them *doublecrossed* moves, because whoever makes them usually has been!

Now Amy's strategy suggests the following policy:

> Make sure there are long chains about and try to force your opponent to be the first to open one.

Try To Get Control...

We'll say that whoever can force her opponent to open a long chain has *control* of the game. Then:

> When you have control, make sure you keep it by politely declining 2 boxes of every long chain except the last.

... And Then Keep It.

The player who has control usually wins decisively when there are several long chains.

So the fight is really about control. How can you make sure of acquiring this valuable commodity? It depends on whether you're playing the odd- or even-numbered turns... .

Dots-and-Boxes—An Introduction

Figure 6. Which is A and which is B?

Throughout this book, we shall always assume that "A" plays the first turn and "B" plays the second. As the game continues, A plays all of the odd-numbered turns; B, the even.

The rule that helps them take control is:

> A tries to make the number of initial dots + doublecrossed moves *odd*.
>
> B tries to make this number *even*.

Be SELFish about Dots + Doublecrosses!

In simple games the number of doublecrosses will be one less than the number of long chains and this rule becomes:

THE LONG CHAIN RULE:

> Try to make the number of initial dots + eventual long chains even if you are A, odd if you are B.

The reason for these rules is that whatever shape board you have on your paper, you'll find that:

$$\begin{array}{l} \text{Number of dots you start with} \\ + \text{ Number of doublecrosses} \\ \hline = \text{Total number of turns in the game.} \end{array}$$

To see this, we assume the initial position consists of m rows of n dots each, so that

$$\text{Number of dots} = m \times n$$
$$\text{Number of horizontal moves} = m \times (n-1)$$
$$\text{Number of vertical moves} = n \times (m-1)$$
$$\text{Number of moves} = 2mn - n - m$$
$$\text{Number of boxes} = (m-1) \times (n-1)$$
$$= mn - n - m + 1$$
$$\text{Number of moves} - \text{Number of boxes} = mn - 1$$
$$\text{Number of completed turns} = \text{Number of moves} + \text{Number of doublecrosses}$$
$$- \text{Number of boxes}$$
$$= mn - 1 + \text{Number of doublecrosses}$$
$$= \text{Number of dots} + \text{Number of doublecrosses} - 1$$

But the last move of the game necessarily completes a box, leaving the final turn incomplete. So adding this turn to the total gives

$$\text{Number of turns} = \text{Number of dots} + \text{Number of doublecrosses}$$

How Long Is "Long"?

We can find the proper definition of long by thinking about Amy's endgame technique. A long chain is one that contains 3 or more squares. This is because whichever edge Babar draws in such a chain, Amy can take all but 2 of the boxes in it, and complete her turn by drawing an edge that does not complete a box. Figure 7 shows this for the 3-square chain. A chain of 2 squares is short because our opponent might insert the middle edge, leaving us with no way of finishing our turn in the same chain. This is called (Figure 8(a)) the *hard-hearted handout*.

When you think you are winning, but are forced to give away a pair of boxes, you should always make a hard-hearted handout, so that your opponent has no option but to accept. If you use a *half-hearted* one (Figure 8(b)) he might reply with a double-dealing move and regain control. But if you're losing, you might

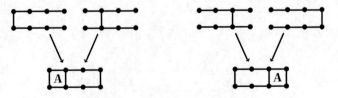

Figure 7. Amy's endgame techniques.

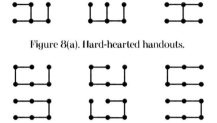

Figure 8(a). Hard-hearted handouts.

Figure 8(b). Half-hearted handouts.

try a half-hearted handout on the Enough Rope Principle. Officially this is a bad move, since your opponent, if he has any sense, will grab both squares. But some mediocre players might blindly blunder both boxes back.

Since the focus of our attention is on *long* chains, we henceforth assume that unless otherwise specified (such as when preceded by the word "short") the term "chain" means "long chain."

The 9-Box Game

The Long Chain Rule ensures that Beth can win the 9-box game. Her basic strategy is to draw 4 spokes as in Figure 9, forcing every long chain to go through the center. Against most children this wins for Beth by at least 6–3, but when Beth plays this strategy Arthur can hold her down to 5–4, perhaps by sacrificing the center square, after which Beth should abandon her spoke strategy. Of course, Beth's real aim is to arrange for just one chain, and she often improves her score by forming this chain in some other way.

Beth usually prefers to put her spokes in squares where another side is already drawn, and she's careful to draw spokes in only one of the two swastika patterns of Figure 9. There usually aren't any double-crossed moves, so that Beth wins at the (16 + 0 =) 16th turn.

Arthur tries to arrange his moves so that some spoke can only be inserted as a sacrifice, and either cuts up the chains as much as possible (maybe with a center sacrifice) or forms two long chains when Beth isn't thinking.

Figure 9. Lucky Charms ward off more than one long chain: Beth puts spokes in Arthur's wheel.

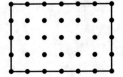
Figure 10. The 4 x 6 Swedish game.

Figure 11. The 4 x 3 Icelandic game.

Other Starting Postions

Swedish children often begin the game with the outside border filled in, as in Figure 10. Another possible starting position, midway between the Swedish and American versions, is the "Icelandic" position shown in Figure 11.

Other Shapes of Board

To beat all your friends on larger square and rectangular boards you'll really need the Long Chain Rule. Remember to count a closed loop of 4 or more cells as an even number of long chains and that each doublecross, no matter who makes it, changes the number of long chains you want. (Think of a doublecross as a long chain that's already been filled in.) It's good tactics to make the long chains as long as possible and avoid closed loops when you can, because you forfeit four boxes when declining a loop. These rules work for all large boards and even for triangular Dots-and-Boxes boards, like that in Figure 12.

Of course, if your opponent is also using the Long Chain Rule, the fight for control might be quite hard. The game of Nimstring (Chapter 6) is what control is all about.

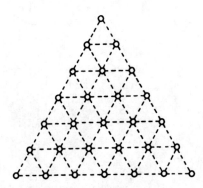
Figure 12. A board with 28 dots and 36 triangular cells.

Chapter 2
Strings-and-Coins

Dots-and-Boxes and Strings-and-Coins

You can play a dual form of Dots-and-Boxes, called Strings-and-Coins, with strings, coins, and scissors. The ends of each piece of string are glued to two different coins or to a coin and the ground (each string has at most one end glued to the ground) and each player in turn cuts a new string. If your cut completely detaches a coin, you pocket it and must then cut another string (if there's one still uncut). The game ends when all coins are detached and the player who pockets the greater number is the winner.

Figure 13 shows the dual of Arthur's and Beth's first game (compare it with Figure 1). It started with 9 coins connected by 24 strings, 12 of them between coins and coins, the other 12 between coins and the ground. We use little arrows for strings that run to the ground. The coins and strings form the nodes and edges of a graph. It's easy to make a graph to correspond to any Dots-and-Boxes position. However, there are lots of graphs which don't correspond to such positions; for example the graph may have cycles of odd length or nodes with more than 4 edges, or the graph may be non-planar. In fact Strings-and-Coins is a generalization of Dots-and-Boxes.

Even though Dots-and-Boxes and Strings-and-Coins are equivalent games, some facts are easier to see from one viewpoint rather than the other. For

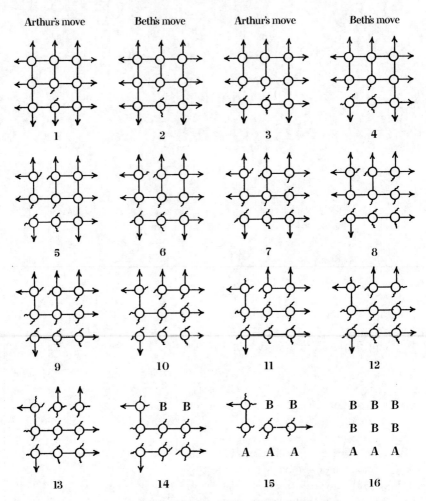

Figure 13. A Strings-and-Coins game—the dual of Figure 1.

example, in any Dots-and-Boxes position, the horizontal and vertical moves in any corner of the board are equivalent. Most students find this fact easier to grasp when viewed from the Strings-and-Coins perspective. To decide whether or not some specific set of boxes forms a chain, most beginners again find the Strings-and-Coins viewpoint to be very helpful.

Chapter 3
Elementary Chain Counting Problems

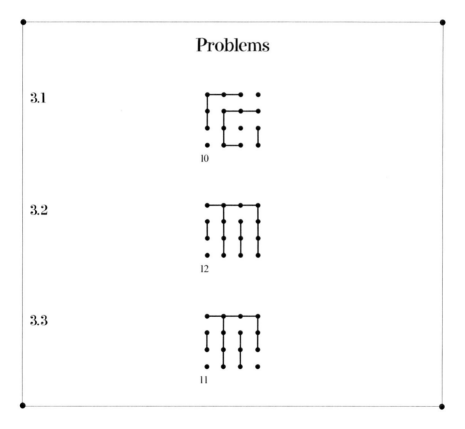

Answers

3.1

11

Ensures 2 chains.

3.2

13

Ensures 2 chains.

3.3

12

Ensures 1 chain.

# Problems

3.4

11

3.5

12

3.6

8

3.7

9

Answers

3.4

12

The dashed move ensures 1 chain.

3.5

13

Decline half-hearted handout to ensure getting the single chain!

3.6

9

Then B keeps control only by sacrificing 2. A eventually wins 5–4.

3.7

10

The only way to stop the side chain is to sacrifice now. Even if A counter-sacrifices, we'll get one chain in the west.

Problems

3.8

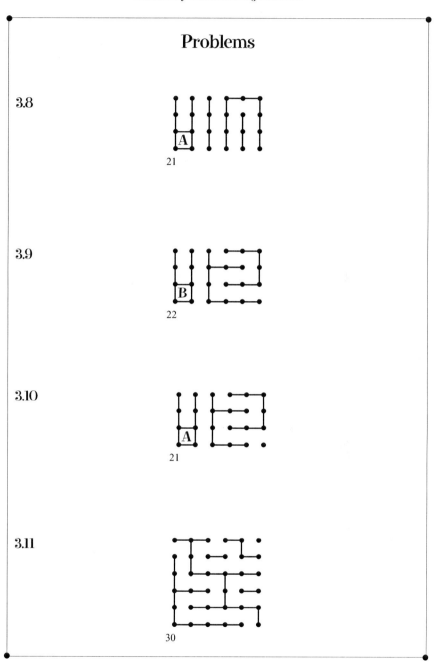

21

3.9

22

3.10

21

3.11

30

Answers

3.8

22

9–6

3.9

23

11–4

3.10

22

13–2

3.11

31

A ensures two chains total by looping the north to prevent a third chain there.

Problems

3.12

28

3.13

22

3.14

13

Answers

3.12

29

3.13

23

Make only 2 chains.

3.14

14

Must prevent a long chain in southwest by sacrificing all 6 boxes there, if necessary.

Problems

3.15

20

3.16

20

Answers

3.15

21

3.16

21

Make 4 chains.

Chapter 4
Advanced Chain Counting

Why Long Is Long

The argument explains why "long" must be defined precisely as follows. We should call a chain *long* if it contains 3 or more coins, because no matter which string of such a chain our opponent might cut, we may take all but 2 of its coins and finish by cutting another string of the chain. We must call a chain of 2 coins *short*, because he might cut the middle string and prevent us from declining those 2 vital coins (the hard-hearted handout). For a similar reason a closed loop of 2 or 3 coins would be called *short* (short loops don't arise in rectangular Dots-

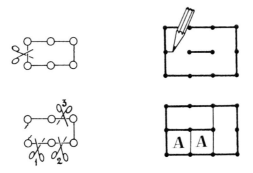

Figure 14. Amy politely declines a long loop.

and-Boxes). However, a loop with at least 4 coins is called *long*, because we can politely decline the last 4 coins no matter which string our opponent cuts. Figure 14(a) shows how to do this on a 6-loop. When your opponent has cut the first string as shown, you only take 2 of the coins and then cut the string in the middle of the remaining 4. Figure 14(b) shows how this corresponds with Amy's way of playing Dots-and-Boxes.

Loony Moves

There are three types of moves which we call "loony":

1). A half-hearted handout

2). A move which offers a long chain (of length at least 3)

3). A move which offers a long loop (of length at least 4)

If our opponent has just made a loony move, then no matter what might be the rest of the position, we can score at least half of the rest of the points!

The proof of this assertion rests on a (non-constructive) strategy-stealing argument. The half-hearted handout is the simplest case; it gives us an immediate choice between two options:

1) We can take the two boxes the opponent has just offered, and go first in the rest of the game, or

2) We can decline the two boxes the opponent has just offered, and thereby make him go first in the rest of the game.

In the latter case, he does best to begin by taking the two boxes himself, and he is in precisely the same situation we were in if we had selected option 1. So, if we knew how (omniscient) gurus would play the rest of the game, we could determine the optimum scores for the first and second players, and then elect to play whichever side could do better.

If opponent's first loony move is to offer us a long chain, of length n, then we can begin by taking $n-2$ of these points, and then decide whether to accept or decline the last two depending on a global analysis which tells us whether we prefer to go first or second on the remaining position. Similarly, if opponent's first loony move is to offer us a long loop, of length n, then we can begin by taking $n-4$ of these points, and then deciding whether to accept or decline the last four points depending on a similar global analysis.

In general, all of the nodes (coins) in any strings and coins position which have valence ≥ 3 (i.e., at least three branches at that node) are called *joints*.

Advanced Chain Counting

The ground may also be viewed as a special joint. Unless they are immediately capturable, all other nodes have valence 2. The branches of the graph can be partitioned into loops and strings. Each string is a path between some pair of joints. Any short loop or short path provides a non-loony move. If all paths and loops are long, then only loony moves are possible, and in this case it is convenient to define the number of long chains as

> Long Chains = Branches − Nodes

This definition gives the right answers when only loops and chains remain. It allows us to count chains in more complicated positions, such as Figure 15. First notice that all moves are loony. Next notice that there are 8 edges and 7 nodes, so the formula says that the number of long chains is $8 - 7 = 1$. By playing the position a few times, we soon realize that it behaves precisely like the sum of one 3-chain followed by one 4-loop.

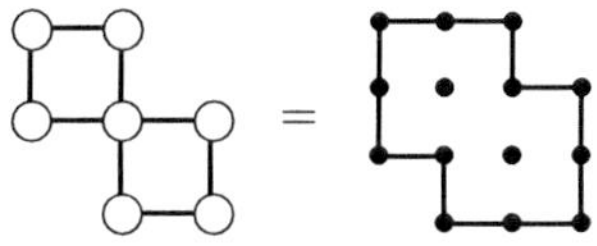

Figure 15. Eight edges − Seven nodes = One chain.

Chains Determined for Some Earlier Positions

In most of the problems of Chapter 2, the locations of the chains could be determined by visual inspection, even though several non-loony moves remained. The key principles are:

> If either player can force the number of chains to be even, then it is even.
>
> *and*
>
> If either player can force the number of chains to be odd, then it is odd.

Examples appear in Figures 16 through 18. In either version of Figure 16, first player can prevent the formation of a long chain by sacrificing the middle box. That cuts the position into two disjoint pieces, neither of which has enough boxes to create a long chain. Likewise, even if the first (i.e., next) player player plays a horizontal move along the bottom edge, second player can still prevent the formation of a long chain by sacrificing the middle square (and possibly one adjacent square). Since *either* player can prevent a long chain, we can say that there are 0 long chains in either version of Figure 16.

a   b

Figure 16. No long chains.

In Figure 17, either player can *create* a long chain. In Figure 17, first player can achieve this goal by playing any horizontal move, except the one in the middle, that does not immediately sacrifice a box. Similarly, second player can also force one chain in Figure 17, no matter what move first play might have played.

In Figure 18, first player can resolve the question of whether or not a long

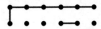

Figure 17. One long chain.

chain will be created, by moving either to Figure 16 or to Figure 17. Thus, in Figure 18, we say that the number of chains is *unresolved*.

In most of the problems in the next chapter, the overall position is comprised of several disjoint subpositions, only one of which has an unresolved number of chains. In such situations, the common winning strategy is to make the move which resolves the total number of chains to whichever parity gives one control.

a   b

Figure 18. Unresolved number of long chains.

Chapter 5
Advanced Chain Counting Problems

Problems

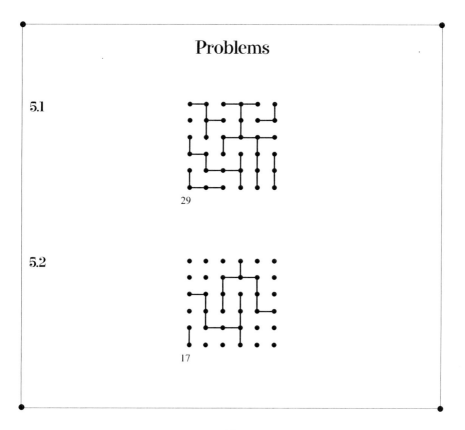

5.1

29

5.2

17

Answers

5.1

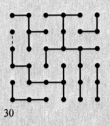

30

Makes 5 chains rather than only 4. After the dashed move, the piece containing 14 boxes has one joint of valence 3. It counts as 2 chains.

5.2

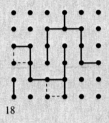

18

Any of the four dashed moves ensures only one chain.

Problems

5.3

17

5.4

24

5.5

20

Answers

5.3

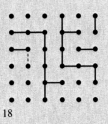

18

Three chains.

5.4

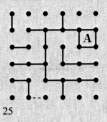

25

Three chains.

5.5

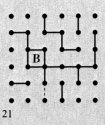

21

Three chains, including the one in the west.

Problems

5.6

28

5.7

26

5.8

20

Answers

5.6

29

5.7

27

This move ensures a chain in the northeast.

5.8

21

Ensures 2 chains.

Problems

5.9

16

5.10

20

5.11

18

Answers

5.9

17

Two chains.

5.10

21

Many moves, including all dashed moves, win big.

5.11

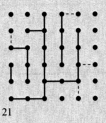

19

The marked move is the best, but there are *many* winning moves.

Problems

5.12

5.13

5.14

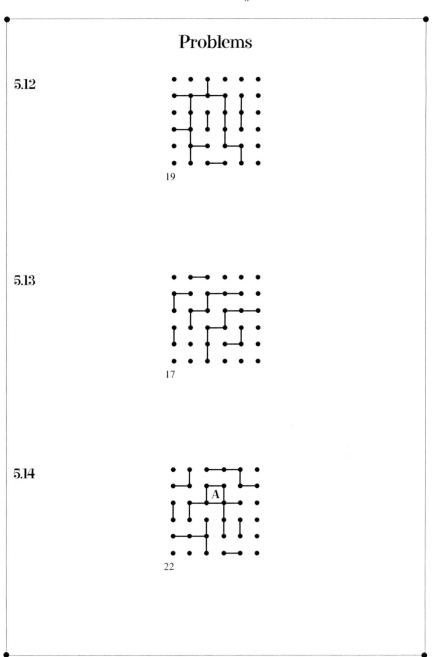

Answers

5.12

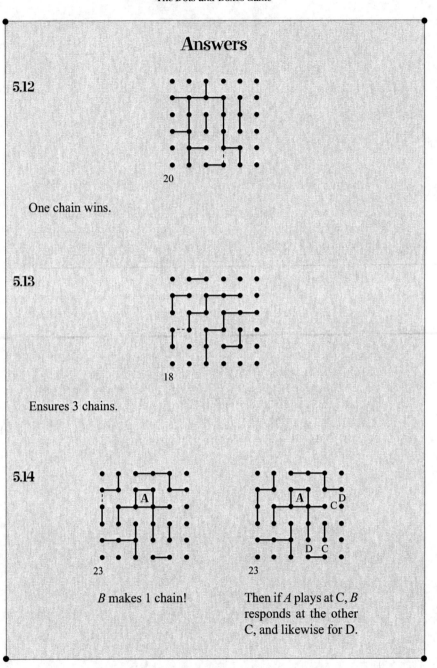

One chain wins.

5.13

Ensures 3 chains.

5.14

B makes 1 chain!

Then if A plays at C, B responds at the other C, and likewise for D.

Problems

5.15

19

5.16

15

5.17

18

Answers

5.15

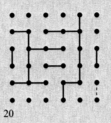
20

Three chains wins.

5.16

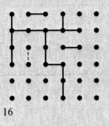
16

Three clear chains, after only 16 moves.

5.17

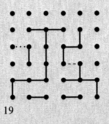
19

A ensures 2 chains by dashed move. If he later gets an opportunity to play the dotted move, it threatens to extend the west into the northwest.

Chapter 6
Nimber Values for Nimstring Graphs

Nimstring

We know that every Dots-and-Boxes position may also be viewed as a Strings-and-Coins position. To advance to the next higher league, we need to master a slightly different game, called Nimstring.

The game of Nimstring is played on exactly the same kind of graphs as Strings-and-Coins, and you make exactly the same move by cutting a string (which gives you an extra turn whenever you detach a coin). In Strings-and-Coins the winner is the player who detaches the larger number of coins, but Nimstring is played instead according to the Normal Play Rule. So, for ordinary Nimstring positions you *lose* when you detach the last coin, for then the rules require you to make a further move when it is impossible to do so. (But a Nimstring graph *may* have a string joining the ground to itself, and if the last move cuts *this* it doesn't detach a coin, and so *wins*.)

Nimstring looks quite different from Strings-and-Coins, but closer investigation shows that Nimstring is in fact a special case of Strings-and-Coins.

> You can't know all about Strings-and-Coins unless you know all about Nimstring!

 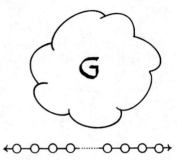

Figure 19 (a). Hard Nimstring problem. (b). This Strings-and-Coins problem is just as hard.

Figure 19 shows the construction which proves this. If G represents an arbitrary Nimstring problem, we add a long chain to it, and consider the resulting Strings-and-Coins game—the long chain should have more coins than G. Because the chain is so long and whoever first cuts a string of it allows his opponent to capture all the coins of the chain on his next turn, both players will try to avoid cutting any string of the chain. Neither player can force his opponent to move on the chain until all the strings of G have been cut. In other words, the only way to win the Strings-and-Coins game of Figure 19(b) is to play a winning game of Nimstring on the graph G.

Figure 20 shows another construction. This time we get the Strings-and-Coins game by adding several long chains and cycles to the Nimstring game G. If these are long enough the winning strategy for the Strings-and-Coins game is then:

> *If your opponent moves in G,* reply in G with a move from the winning Nimstring strategy.
>
> *If he moves in a long chain,* take *all but two* coins of that chain, leaving just the string which joins them.
>
> *If he moves in a long cycle,* take *all but four* coins of the cycle, leaving them as *two pairs* each joined by a string.

This strategy gives you all but 2 coins of each long chain and all but 4 of each long cycle, so it will win for you if the total number of nodes in the added chains and cycles exceeds

$$\begin{aligned}&\text{(the number of nodes in } G\text{)}\\+\ 4\ \times\ &\text{(the number of added long chains)}\\+\ 8\ \times\ &\text{(the number of added long cycles)}.\end{aligned}$$

Nimber Values for Nimstring Graphs

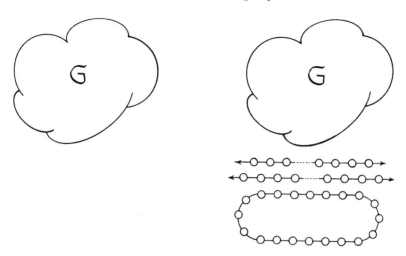

Figure 20 (a). Another Nimstring game; (b). A Corresponding Strings-and-Coin game.

In practice the Nimstring position will often contain (potential) long chains of its own, so that the strategy is of wider application. Recall that the "all but 2" principle was used by Amy against Babar (Figure 2). Many well-played games of Dots-and-Boxes are played like the corresponding Nimstring games, except at the very end. The last long chain in a Nimstring game is treated like any other: the winner takes all but the last 2 coins, which he gives to the loser by a hard-hearted handout. For the last chain in Dots-and-Boxes, of course, winner takes all!

Well-played games of Dots-and-Boxes frequently lead to the duals of positions like those in Figure 20(b). Most of the coins are in the long chains and loops, and the winner is whoever can force his opponent to cut the first string in one of those. It seems to be very often the case that the winning strategy for Nimstring also gives the winning strategy for Strings-and-Coins. There are many other graphs than those satisfying the conditions of Figure 20(b) for which this can be proved to happen. To win a game of Dots-and-Boxes or Strings-and-Coins, you should try to win the corresponding game of Nimstring and at the same time arrange that there are some fairly long chains about. In the rest of this chapter we'll teach you how to become an expert at Nimstring.

To Take or Not To Take a Coin in Nimstring

A coin which has only a single string attached is *capturable*. Whenever there's a capturable coin the next player has the option of removing the corresponding branch, thereby detaching the coin and getting another (complimentary) move. For some graphs this is the best move; for others, including one of those encountered by Amy in the game of Figure 2, the winning strategy is to refuse to detach the coin. As you might guess, the decision as to whether it's better to take a coin or decline it often depends on the entire graph. However, a great deal can be deduced by examining only local properties of the graph near the capturable coin.

Any capturable coin must look like one of the six possibilities in Figure 21. The string from the capturable coin goes either to the ground (Figure 21(a)) or to another coin. If to another coin, the number of strings there is either one (Figure 21(b)), two (Figures 21(c), (e), and (f)), or three or more (Figure 21(d)). If there are two strings, the second goes either to another capturable coin (Figure 21(c)), or to the ground (Figure 21(e)), or to a coin with two or more strings (Figure 21(f)). In each of the six cases the cloud contains all the coins and strings not regarded as near enough to the capturable coin. The dotted lines in Figures 21(d) and (f) are possible additional strings which may or may not be present.

We claim that in the first four cases (Figures 21(a)–(d)) the player to move might as well cut string X and capture the coin, and in Figure 21(c), he might as well continue by cutting string Y, taking two more coins. For suppose you

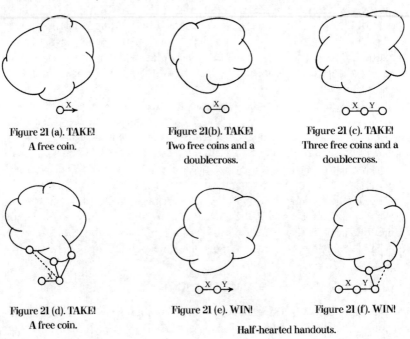

Figure 21 (a). TAKE!
A free coin.

Figure 21(b). TAKE!
Two free coins and a doublecross.

Figure 21 (c). TAKE!
Three free coins and a doublecross.

Figure 21 (d). TAKE!
A free coin.

Figure 21 (e). WIN!
Half-hearted handouts.

Figure 21 (f). WIN!

have a winning strategy starting from one of these graphs. If this tells you to complete your first turn by cutting only certain unlettered strings, then your opponent has the option of beginning his turn by cutting the lettered ones. But the same position will be reached if instead you first cut all the lettered strings and then cut the same unlettered ones as before. If there's any winning strategy at all, starting from these four cases, there's one that begins by cutting the lettered strings. So there's no loss in generality in supposing that a good player will TAKE a capturable coin of one of the four types in Figures 21(a)–(d).

The other two positions (Figures 21(e) and (f)) are much more interesting. If it's your turn to move in one of these two cases, you can either detach the capturable coin by cutting string X, or decline to take it by cutting string Y. No

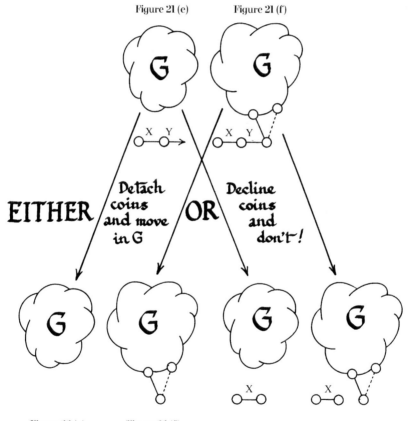

Figure 22. Strategy stealing after a half-hearted handout.

matter what the rest of the graph might be, one or other of these two moves will WIN. But you might need to look at the whole graph to decide whether your winning strategy begins by cutting string X or string Y!

This somewhat surprising result can be provided by a cunning use of Strategy Stealing (Figure 22).

We ask, for the games of Figures 19(e) and (f):

who wins the smaller game G consisting of just the unlettered strings (Figures 22(e) and 22(f)?

This is either the player who has to move from G or the player who doesn't. Whoever this fortunate player is, you should arrange to steal his strategy. If the player to move from G can win, then when playing from Figure 21(e) or (f) you should start by cutting string X (which detaches a coin, so you continue), then cut string Y (detaching another coin, so you continue again) and then begin the game on G, which of course you will play according to the winning strategy for the first player. On the other hand, if there's no winning move for the first player from G then, starting from Figure 21(e) or (f), you should finish your turn immediately by cutting string Y and so force your opponent to start the game G (he might as well start by cutting string X; if he doesn't, you will later.

The fact that the declining move forfeits 2 coins to your opponent makes no difference in Nimstring, where the winner is determined by the last move. In Strings-and-Coins (and Dots-and-Boxes) it might matter, but is unlikely to when there are long chains about.

Nimber Values for Nimstring Graphs

Every nimstring position can be characterized by a single entity, called its *value*. This value can be a "nimber" corresponding to any of the ordinary nonnegative integers, or a special value, $\mathbb{D}$, called "loony", which in certain ways behaves like a nimber corresponding to ∞. Figure 23, shows an example of a loony nimstring position.

Figure 23. A loony nimstring (or dots-and-boxes) problem.

The Mex Function

The value of a nimstring position can be computed by a simple recursion involving a function called the mex (an abbreviation for "minimal excludant"). The mex of any finite set of nonnegative integers is the least nonnegative integer not in the set.

For example,

$$\text{mex}(5, 18) = 0$$
$$\text{mex}(0, 1, 2, 3, 5, 7) = 4$$
$$\text{mex}() = 0$$

The last equation states that the mex of the empty set is 0.

Positions of the type shown in Figures 21(e) and (f) have a special value, $\mathcal{D}$. When this value appears as an argument of the mex function, it can be ignored:

$$\text{mex}(\mathcal{D}, a, b, c) = \text{mex}(a, b, c)$$

The Recursion for Nimber Values

Here is the recursion to calculate the values of an arbitrary Nimstring Graph:

> The value of the empty graph (without any strings) is 0.
>
> The value of a graph with a capturable coin of one of the four types in Figures 21(a)–(d) is equal to that of the subgraph obtained by removing the capturable coin(s) and its string(s).
>
> The value of a graph with a capturable coin of one of the two types in Figures 21(e) and (f) is $\mathcal{D}$.
>
> The value of a graph with no capturable coins is found from the values of the graphs left after cutting single strings by using the Mex Rule.

In the jargon of combinatorial game theory, Nimstring belongs to a category of games called "Impartial". The assiduous reader can find many references to such games in the literature, including "Fair Game" by Richard Guy [4], and Chapters 4, 14, 15, 16 of "Winning Ways" [1]. It is conventional to precede all nimber values, except 0 and $\mathcal{D}$, by the prefix "*". Nimber values for several simple nimstring positions are worked out in Figure 24. The value of each position in this figure appears just below it. Within each position, the value of the position obtained by deleting any single branch is shown (without the * prefix) on the branch. In particular, one interesting example is the 2 × 2 Icelandic game, which often appears as one region among many others in a much larger position. The equivalent nimstring graph appears at the bottom right of Figure 24, where we find that the eight branches that might be removed yield values (ordered from left to right, top row first) of *3, *1, *3, *1, 0, *3, 0, *3, from which the recursion gives the nimber value of the 2 × 2 Icelandic game as

$$\mathrm{mex}(3, 1, 3, 1, 0, 3, 0, 3) = 2.$$

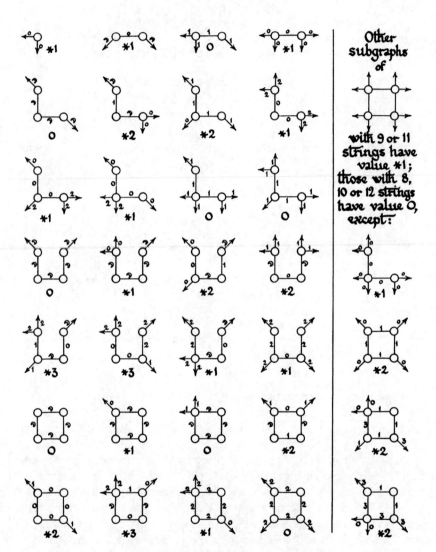

Figure 24. Working out values for nimstring graphs.

(Optional) Nimstring Problems

6.1 $G =$ [figure] $= ?$

7

6.2 $G =$ [figure] $= ?$

6

Answers

6.1

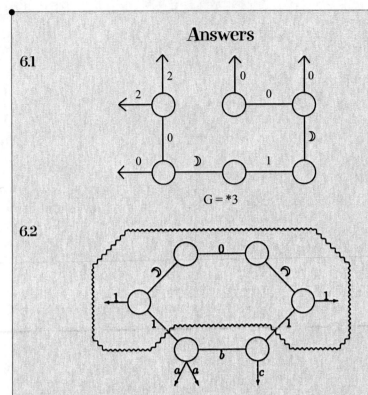

$G = *3$

6.2

Let G_a, G_{ab}, etc., denote the graphs with the subscripted branches deleted. Obviously $G = \text{mex}(0, 1, G_a, G_b, G_c)$.

The five non-loony moves within the cloud force the number of long chains to be 0 or 1 according to the label shown. This statement remains true if any of the following combinations of branches are removed:

$$a \text{ or } b \text{ or } ab.$$

Hence $G_{ab} = \text{mex}(0, 1) = *2$, and $G_a = G_b = *3$.

But we must evaluate $G_{aa} = G_{ac}$, G_{bc}, and G_c separately. From G_{ac} or G_{bc}, the second player can force one chain with no other branches. So $G_{ac} = G_{bc} = 0$. That done, we obtain

$$G_c = \text{mex}(0, 1, G_{ac}, G_b) = *2$$

Whence

$$G = \text{mex}(0, 1, 2, 3) = *4.$$

Who Wins Nimstring?

The following rule shows us how these nimber values allow us to determine who can win from any position at Nimstring:

> If a Nimstring position has any value other than 0, then the next player to move can win by moving to a position of value 0.
>
> If a Nimstring position has value 0, then the player who has just made the prior move can win.

In particular, the first player can win a game of Nimstring played on the small 2 × 2 Icelandic starting position.

Who Wins Sums of Nimstring Positions?

There is a rule which allows us to calculate the value of any SUM of nimstring positions directly from the values of the individual summands. This rule was discovered in 1901 by a Harvard mathematician named Bouton:

The nimber addition rule:

    1. Write the nimber value of each summand in binary.

    2. Form the XOR (exclusive-or) of these binary numbers.

    3. That XOR result is the nimsum.

Loony values follow special rules:

$$\text{For any } n, \quad \mathcal{D} + *n = \mathcal{D}, \quad \text{and } \mathcal{D} + \mathcal{D} = \mathcal{D}$$

This is a powerful result, which enables us to analyze many positions with much less effort than would otherwise be required. For example, consider the 4 × 5 Dots and Boxes position in Figure 25. In this position, 16 moves have been played, and there are 33 possibilities of the next move. A direct attempt to compute the nimber value of Figure 25 recursively would eventually entail looking at all 2^{33} subpositions. Fortunately, the corresponding nimstring graph is the sum of three positions. In Figure 25, the "North" comprises the first two rows. The "Southeast" is the 2 × 2 Icelandic corner, and the "Southwest" comprises 6 boxes which appeared in Problem 6.2. Because so many of the moves in the North are loony, its value is easily calculated to be *1. Since we already

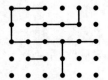

Figure 25. A 4 × 5 position including Problem 6.2 and a 2 × 2 Icelandic corner.

know the values of the southeast and southwest, we can apply the nimber addition rule as follows:

$$
\begin{array}{lll}
\text{North} & = *1 & = 001 \\
\text{Southeast} & = *2 & = 010 \\
\text{Southwest} & = *4 & = 100 \\
\hline
\text{Figure 25} & & = 111 = *7
\end{array}
$$

Since *7 is nonzero, there is a winning move. To find it, we first observe that the leading nonzero bit in *7 is the fours bit. In the XOR calculation, this bit came from a nonzero bit in the Southwest. So we can change the XOR value from *7 to 0 by changing the Southwest to Southwest':

$$
\begin{array}{lll}
\text{North} & = *1 & = 001 \\
\text{Southeast} & = *2 & = 010 \\
\text{Southwest}' & = *4 & = 0?? \\
\hline
\text{Figure 25} & & = 000
\end{array}
$$

In this case, the required value of Southwest' will have to be 011, or *3. A review of the answer to Problem 6.2 reveals that a winning Nimstring move from Figure 25 lies in the southwest corner of the board.

Nimbers and Chain-Counts

Any positions in which the number of long chains is resolved has nimber which is either 0 or 1, accordingly as the parity of the number of long chains, plus the number of nodes (boxes) plus the number of edges (moves) is even or odd. Positions such as Problems 6.1 and 6.2, in which the parity of the number of long chains is unresolved, but can be resolved by the first player, have "big" nimber, ≥ 2. This corresponds to the fact that the mex function can attain a value ≥ 2 if and only if its inputs include *both* 0 and 1.

In prior chapters of this book, we found that chain-counts were often very effective tools for finding winning moves in Dots-and-Boxes. When all-but-one regions of the board contain a known number of long chains, and one region remains unresolved, then, if possible, the first player should resolve the number of chains in that region to attain whichever total parity he desires. These situations correspond to the cases in which all but one region of the board have nimber values which are zero or one, and the remaining region has a big nimber (≥ 2). In that case, a winning move is clearly to reduce the big nimber to 0 or 1, as needed to bring the total nimber value to 0.

Chain counting proves inadequate when several regions of the board contain unresolved chain-counts. In particular, if exactly two regions of the board are unresolved, then it is a mistake to resolve either, because your opponent can then resolve the other in his favor. Fortunately, nimbers are more powerful than chain-counts, because they successfully categorize the regions in which the chain count is unresolved.

Nevertheless, chain-counts are very convenient. As we have seen in the problems of Chapters 3 and 5, chain counting is often much easier than the recursive computations which define nimber values in general. This is closely related to the fact that, if a position has nimber value 0 or 1, it is sufficient to know whether each of its followers has nimber value 0, 1, or BIG, and one doesn't need to know the precise values of the BIGs. Hence, expert nimstring players need to be facile with *both* chain-counts *and* nimbers.

Students who attempt to relate these two approaches will eventually discover that there is another interesting class of positions, in which the number of chains can be resolved by the *second* player. One example of this occurs after a single corner move is played in the Icelandic 2×2 corner, yielding the position shown in Figure 26. In this Figure, Bertha, if she chooses, can force an odd number of long chains. Or, if she chooses, she might instead force an even number of long chains. Since Arthur has so little influence on this nimstring position, it is not surprising that he cannot win if this small position constitutes the entire game. Since Figure 26 (if played alone) is a win for the second player, its nimstring value must be 0. In the discussion on page 25, we saw that when the position is loony (or, more generally, if the nimstring value is zero) it is convenient to anticipate the parity of the number of long chains from the formula

$$\text{Long Chains} \equiv \text{Branches} - \text{Nodes} \pmod 2$$

In Figure 26, there are 4 boxes and 7 moves remaining. So, we should infer that the number of chains is odd. What this means is that, even in the context of a much-larger game, we can make the canonical restriction that, whenever

Figure 26. The 2 × 2 Icelandic game after one move.

a nimstring player plays on Figure 26, his opponent will always respond immediately with another move in the same region which ensures that there is a single long chain there. Either player who has a winning strategy can still win if he accepts the canonical restriction, even if his opponent ignores it. So the canonical restriction provides some simplification without any loss of rigor.

Thus, an experienced player might offer the following justification of the fact that the 2 × 2 Icelandic corner has a "big" nimber:

> 1) First player can force 0 chains by immediately sacrificing the interior corner.
>
> 2) First player can force one chain by playing immediately in the exterior corner.
>
> 3) Therefore, the parity of the number of long chains is unresolved, and the nimber is BIG. (i.e., ≥ 2).

The second of these assertions is surely the most subtle.

On-Line Nimstring Program

Freddy Mang wrote a computer program which computes nimstring values of Dots-and-Boxes positions. It is convenient to use, and is available on the web at http://www-cad.eecs.berkeley.edu/~fmang/nimstring/index.html. (For updated references see www.akpeters.com.)

Chapter 7
Elementary Problems with Nimbers

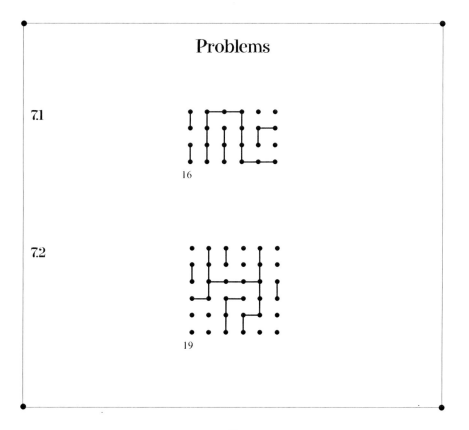

Answers

7.1

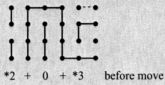

*2 + 0 + *3   before move
*2 + 0 + *2   after

7.2

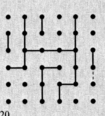

20   gives these nimbers

Problems

7.3

19

7.4

15

7.5

19

Answers

7.3

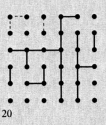

20

Any of the three dashed moves forces a chain in the northwest (and so do two others).

7.4

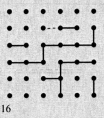

16

Since the southeast corner counts as one chain, the move shown gives a total of 3.

7.5

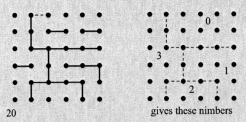

20   gives these nimbers

In the northeast, the first player can force a chain but he cannot prevent a chain, so the original northeast is *. Arthur too has several winning moves, including the dashed move shown or a move which changes the west from *3 to *2.

Problems

7.6 Can *Second* player force an even number of chains in this position?

(This problem is used in the sequel).

7.7 Verify that the nimstring value of this position is * (≥ 5):

7.8

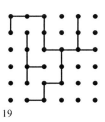

19

Answers

7.6 YES. Any move can be answered to one of the following:

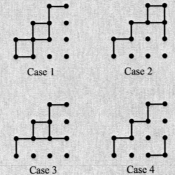

7.7

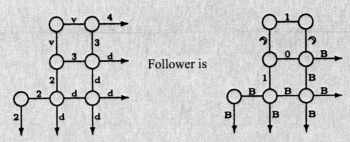

d means odd # chains
v means even # chains

Follower is

Each B means ≥ 2
(Different Bs need not be equal.)

7.8

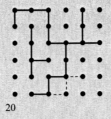

Any dashed move gives *2 in southeast corner and wins.

Chapter 8
More about Nimstring, Arrays, Mutations, Vines, etc.

All Long Chains Are the Same

Look at the various positions of Figure 27, in which the clouds all conceal exactly the same thing, and the necklaces that hang from them all have at least three beads. The graphs all behave the same way in Nimstring because all the visible edges will always be loony moves.

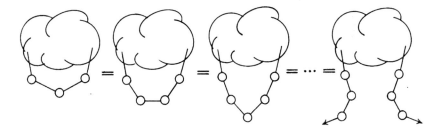

Figure 27. Three or more's a crowd.

> Provided a chain has 3 or more nodes along it, the exact number doesn't make any difference to the nimstring value.

This makes it handy to have a special notation for long chains:

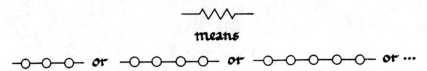

Which Mutations Are Harmless?

More generally we can put in or take out some beads on any Nimstring graph G to obtain mutations of the graph (a bead, of course, is a node with just 2 edges). Figure 28 shows a graph G and two mutations, H and K.

We'll use the word stop to mean either an arrowhead where the graph goes to ground (an end) or any of the nodes which have 3 or more edges (the joints). A path between two stops is long if it passes through 3 or more intermediate nodes, short otherwise. Mutation usually affects the value, but there are a lot of harmless mutations that don't:

> A mutation between two graphs will certainly be harmless if every short path between stops in either graph corresponds to a short path in the other.

The Harmless Mutation Theorem.

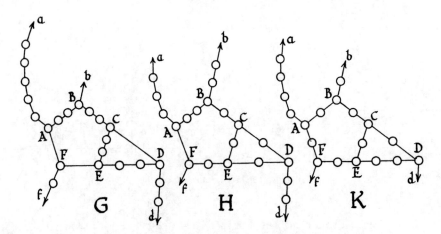

Figure 28. A graph, a harmless mutation and a killing one.

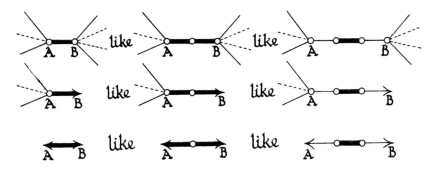

Figure 29. Like moves in harmless mutations.

In Figure 28, H is a harmless mutation of G, since the only short paths are AE, Af, Ef, and the ones other than Aa that don't pass through a stop. But AE is long in K, and Cd is short, so this mutation is not covered by our theorem. In fact G and H have value *2, while K has value 0.

When G and H are related by a harmless mutation you just play H like G. A non-loony move must cut some string of a short chain between two points A and B that were stops at least until the move was made. A and B must have been stops in the original graph and we can find a similar non-loony move in the mutated graph because the distance between A and B will be short. (Figure 29).

We can strengthen our Mutation Theorem a little:

> If the path between two stops passes through one end of a long chain, you needn't worry about the length of the path.

(For in a graph like Figure 30—in which A or B might have been ends—AB won't become a chain unless someone makes a loony move cutting the long chain ending at C.)

Chopping and Changing

There are lots of more drastic changes we can make to Nimstring graphs without affecting their values; for instance:

> Long chains snap!

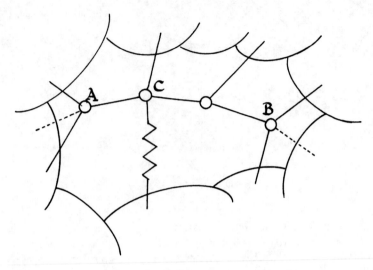

Figure 30. The path AB passes a long chain at C.

This was hinted at in Figure 27, and Figure 31(a) shows how it's written in our long chain notation.

The remaining equivalencies of Figure 31 are more interesting. The middle equivalence of Figure 31(b) is particularly useful (the left equivalence is a long chain that snaps). It asserts that when an edge runs to a node from which two long chains emanate, then this edge may be replaced by an edge running directly to the ground. More generally, when two long chains are attached to a node, all other edges ending at this node may be replaced by edges running to the ground (Figure 31(c)).

The idea of the proof is that a node at the end of two long chains can't be captured until after someone concedes the game by making a loony move. We can apply the equality between the first and last parts of Figure 31(b) to every branch that runs to the ground, so as to eliminate ground branches from every graph. But usually it's more convenient to use it in the other direction, eliminating many branches and nodes by introducing new ends. Sometimes, as in Figure 31(d), this gives rise to a branch joining the ground to itself (a 0 by 1 game of Dots-and-Boxes!); such a branch contributes *1 to the value.

The equality between the first three parts of Figure 31(e) follows from the Harmless Mutation Theorem, but that between these and the last three doesn't, because some short chains have become long. The letters label corresponding moves, and the $\mathcal{D}$'s show moves which we should ignore. Figures 31(b), (d), (f), and (g) show that we can sometimes eliminate circuits from our graphs—the last

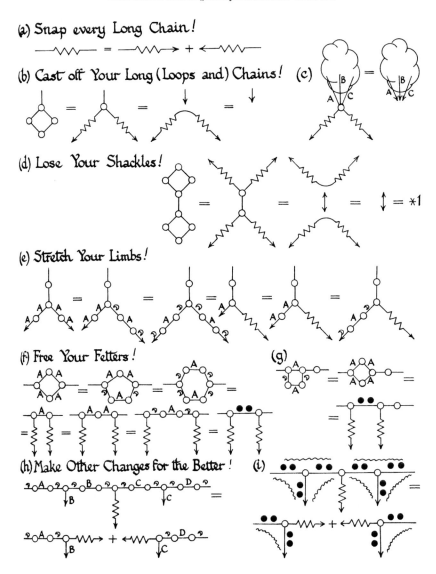

Figure 31. Some useful nimstring equivalences.

diagram of Figure 31(f) is our shorthand notation for any of the previous three, which are harmless mutations of each other. Figure 31(h) has many variants, abbreviated in Figure 31(i) (using the notation of Figure 32).

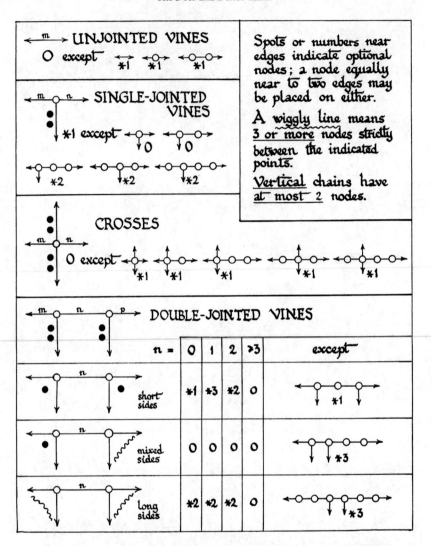

Figure 32. Noteworthy nimstring nimbers.

Vines

A vine is a Nimstring graph without circuits or capturable nodes in which all the joints lie on a single long path (the stem) and each joint belongs to just 3 edges. The chain joining an end to its nearest joint is called a tendril, so a single-jointed vine has 3 tendrils. Vines with more joints have 2 tendrils at their endmost joints and just 1 at intermediate ones. If the distance between two neighboring joints is long, the vine decomposes into two smaller ones because long chains snap, so we can suppose such distances short, if we like.

The nimber values of all vines having at most two joints are shown in Figure 32. A Twopins-vine is one whose every distance between non-neighboring stops (which may be either ends or joints) is long. Such vines have the interesting property that any single non-loony move soon yields a new position which is the sum of simpler twopins-vines.

A non-loony tendril move will necessarily create a long chain along the stem of a twopins-vine. If there are other tendrils on both sides of the one just played, then this long chain can be snapped to break the position into a pair of twopins-vines, whose total number of tendrils is one less than the original twopins-vine. If one instead plays a non-loony stem move in the interior of a twopins-vine, then the twopins-vine is broken into two pieces, each of which is a smaller twopins-vine. The adjacent tendrils on both sides of this stem become ends of the new twopins-vine, and total number of tendrils is two less than the original.

There are two interesting special cases of twopins-vines for which an unusually detailed and complete analysis is known. These are Dawson's-vines (all of whose tendrils are long) and Kayles-vines (all of whose tendrils are short). The nimber value of Dawson's-vine with n joints satisfies the recursion

$$D_n = \mathop{\mathrm{mex}}_{k} (D_k + D_{n-k-2})$$

This recursion is self-starting, with D_0 and D_1 being the mex of the empty set, which is 0. Then

$$D_2 = \mathrm{mex}\, (D_0 + D_0) = \mathrm{mex}\, (0) = 1$$
$$D_3 = \mathrm{mex}\, (D_0 + D_1) = \mathrm{mex}\, (0) = 1$$
$$D_4 = \mathrm{mex}\, (D_0 + D_2,\ D_1 + D_1)$$
$$= \mathrm{mex}\, (1, 0) = 2$$

where we use the nimsum addition rule to compute

$$D_0 + D_2 = 0 + {*}1 = {*}1$$

and

$$D_1 + D_1 = 0 + 0 = 0$$

Similarly,

$$D_5 = \text{mex} (D_0 + D_3, D_1 + D_2)$$
$$= \text{mex} (1, 1) = 0$$
$$D_6 = \text{mex} (D_0 + D_4, D_1 + D_3, D_2 + D_2)$$
$$= \text{mex} (2, 1, 0) = 3 \text{ etc.}$$

After some exceptional values shown in boldface in the following table, these values eventually settle down into a pattern with period 34:

n	0	1	2	3	4	5	6	7	8	9	11	13	15	17	19	21	23	25	27	29	31	33
D_n	0	0	1	1	2	0	3	1	1	0	3 3	2 2	**4 0**	5 2	2 3	3 0	1 1	3 0	2 1	1 0	4 5	2 7
D_{n+34}	**4**	0	1	1	2	0	3	1	1	0	3 3	2 2	**4 4**	**5 5**	2 3	3 0	1 1	3 0	2 1	1 0	4 5	**3** 7
D_{n+68}	**4**	**8**	1	1	2	0	3	1	1	0	3 3	2 2	**4 4**	**5 5**	**9** 3	3 0	1 1	3 0	2	...		

According to this table, the last exceptional value is $D_{34+18} = 2 \neq D_{64+18} = 9$. And in fact, for all $k > 82$, $D_k = D_{k-34}$.

The nimber values of Kayles-vines satisfy a similar recursion:

$$K_n = \text{mex}_j (K_j + K_{n-j-1}, K_j + K_{n-j-2})$$

whose solutions are:

n	0	1	2	3	4	5	6	7	8	9	10	11	12	13	14	15	16	17	18	19	20	21	22	23
K_n	0	1	2	3	1	4	3	2	1	4	2	6	4	1	2	7	1	4	3	2	1	4	6	7
K_{n+24}	4	1	2	**8**	**5**	4	7	2	1	**8**	**6**	**7**	4	1	2	3	1	4	7	2	1	**8**	**2**	7
K_{n+48}	4	1	2	**8**	1	4	7	2	1	4	**2**	**7**	4	1	2	**8**	1	4	7	2	1	**8**	6	7
K_{n+72}	4	1	2	**8**	1	4	7	2	1	**8**	2	7	4	1	2	**8**	1	4	7	2	1	**8**	2	7

Here again, there exceptional values, which are shown in boldface, die out, and the sequence becomes purely periodic. This time the period is 12, and the last exceptional value is $K_{70} = 6 \neq K_{82} = 2$. But for all $n \geq 83$, $K_n = K_{n-12}$.

All twopins-vines are decomposing in the sense that when any branch of the vine is removed the new vine decomposes—often by snapping a chain—into two smaller ones. Some other vines, including that of Figure 33(b), are decomposing in the same sense. It is rather straightforward to compute the value of a decomposing vine from the values of those of its subvines which include all of a

VINES

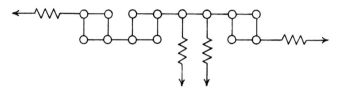

(a) A graph equivalent to the Dawson's-vine D_8.

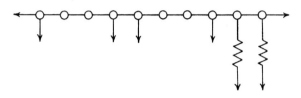

(b) A decomposing vine.

Figure 32. Two uses of Figure 31.

consecutive sequence of the original tendrils. Since the number of such subvines is proportional only to the square of the number of tendrils this idea is feasible for quite long decomposing vines, and can easily be implemented on a computer.

More information about the nimber values of twopins-vines can be found in Winning Ways [1], which also contains much material on other impartial games closely related to Kayles-vines and Dawson's-vines.

Chapter 9
Advanced Nimstring Problems

Problems

9.1
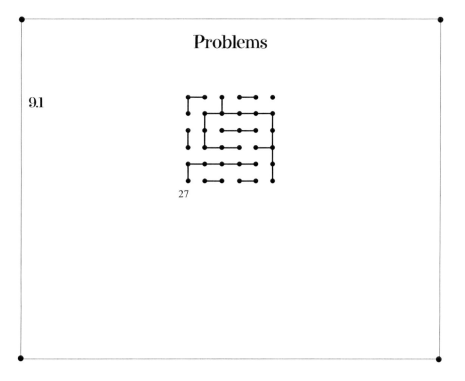
27

Answers

9.1

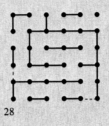

28

In the top right corner, we have 3 boxes that can be made into either one long chain or zero long chains; and any move there will force the situation one way or the other. Hence its nimber is 2. In the middle is an 8-box loop, with nimber 0. The rest of the game is a *Kayles-vine* with 5 tendrils. The nimbers for Kayles are:

n:	0	1	2	3	4	5 ...
nimber:	0	1	2	3	1	4 ...

So the nimber of the whole game is now *2 + *4 = *6. We must change the 4 to a 2. As seen from the above list of nimbers for Kayles, this can be accomplished only by changing the 5-tendrils into 1-tendril and 3-tendrils. This requires a tendril move rather than a stem move, and there are precisely two such moves, marked above.

Since all long chains will eventually be ours, we maximize our score by marking them as long as possible. The move on the left side creates a chain of length 4, while the move at the bottom creates one chain only of length 3, so there is a preference for the former.

Problems

9.2

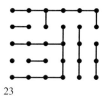

23

9.3

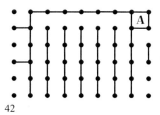

42

Answers

9.2

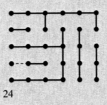

24

The twopins-vine has value 0, and the move shown reduces the lower left corner from *2 to 0.

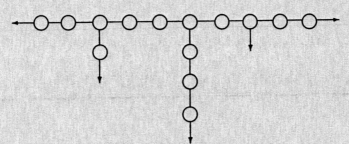

9.3

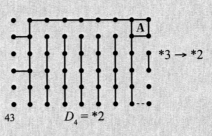

43 $D_4 = {*}2$

*3 → *2

Problems

9.4

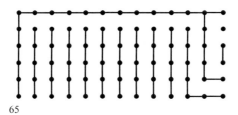

65

9.5

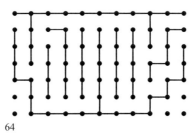

64

Answers

9.4

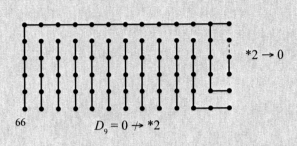

66     $D_9 = 0 \not\to *2$

$*2 \to 0$

9.5

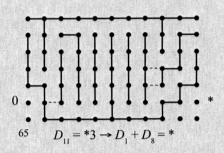

0   *

65     $D_{11} = *3 \to D_1 + D_8 = *$

All three winning moves are dashed.

Chapter 10
Playing Dots-and-Boxes with Very Close Scores

Nimstring Incentives

In the game of Nimstring, the amount by which a move changes the nimstring value is called its "nimstring incentive". A very common incentive is *1, which is often abbreviated more succinctly as *. Any nonzero nimber can be an incentive. For example, suppose a Nimstring game is played on the sum of three disconnected components, X, Y, and Z, whose respective nimstring values are 0, *, and *3. The total position is

$$G = X + Y + Z$$
$$= 0 + * + *3$$
$$= *2$$

Any winning move from G must change the value from *2 to 0, so it must have incentive *2. One such move would change the value of Z from *3 to *. Another such move might change the value of Y from * to *3. Since the nimber value of Y is 1, we know that the nimber values of its followers, Y′, satisfy

$$1 = \text{mex}(Y')$$

so Y′ necessarily includes 0 and excludes 1. Y′ might or might not include the nimber 3.

Canonical Play

It often happens that one player has several choices of winning moves. To simplify one's strategizing, it may be helpful to narrow this range of choices by arbitrarily imposing some additional restrictions called "canonical" rules. This term is reserved for rules which satisfy an important property:

> In any position, if you have any move which wins against a perfect opponent, then you have a canonical move which wins.

Alternatively, if we wish to expand our list of winning options, then we should also consider non-canonical moves.

Non-Canonical Nimstring Moves Might Win

In the game of nimstring, a canonical move is one which lowers the nimber-value of the region in which it is made. Any move which increases the nimber value of the region in which it is made is non-canonical. We saw in Chapter 6 that if there is any winning nimstring move, there is a canonical winning nimstring move.

In particular, if several components have nimber values which are 0 or 1, and only one component of the position has a value which is a "big" number ≥ 2, then there is necessarily a canonical winning nimstring move on the big component. However, there might also be additional winning nimstring moves on one or more of components whose value is 0 or *.

Moves which have the same nimstring incentives can have quite different effects on the string-and-coins score. When this score is close, winning the strings-and-coins game can depend on choosing which of these winning nimstring moves is the best for strings-and-coins.

Canonical Strings-and-Coins Moves

It is also possible to simplify the choice among strings-and-coins moves by introducing "canonical" restrictions. For example, we decree that

> A half-hearted handout is NOT canonical.

So, a canonical player will always prefer a hard-hearted handout to a half-hearted one. The latter gives the opponent more options than the former, and, against an expert opponent, these options can only serve to help him.

There is no assurance that a noncanonical move will lose or that a canonical move will win. Rather, we claim only that if there is a winning non-canonical move from any given position, then there is also a winning canonical move.

To Take or Not to Take at Strings-and-Coins

When playing Dots-and-Boxes, or the equivalent game of Strings-and-Coins, most of the conditions under which a good nimstring player would take a point remain valid.

In particular,

> A canonical strings-and-coins player will *take* a coin in any of the conditions shown in Figures 21a, 21b 21c, and 21d.
>
> He will also take **all-but-two** of any long chain which he is offered, and **all-but-four** of any long loop which he is offered.

However, when he is down to the last four coins in any long loop, or the last two coins in any long chain, (e.g., Figure 21e), the canonical player will *choose*. Sometimes his decision will differ from that of the Nimstring player. Whereas the nimstring player will select the option which ensures herself the last move in the overall game, the canonical Strings-and-Coins player will select an option which ensures himself at least half of the coins which remain untaken at the time the decision is made. Since a canonical *guru* (who is able to analyze the entire game flawlessly) will always make such a decision correctly, it follows that in a game in which the total number of points is odd,

> When played against a canonical guru, any loony move is an losing move *unless* you are far enough ahead in score that you remain ahead when your opponent makes his choice.

If the total number of points in the entire game is even, then possibly you might be able to get a tie against a canonical guru by playing a loony move when you are ahead only enough to ensure that the score is tied when he makes his choice.

In particular, suppose you have scored k more boxes than your opponent so far in a Dots-and-Boxes game which began with an odd × odd number of boxes (and an even × even number of dots). Suppose that you next play a loony move, offering him an n-chain, where $n \geq 3$. Then he will begin by taking $n - 2$ boxes. When he makes his decision about whether or not to take the last two boxes, your lead will be $k + 2 - n$. If this is negative, the canonical guru will certainly win. He will also win if $n = k + 2$, because in that case, there will still be an odd number of points available when he makes his choice, and he'll find the choice which ensures that he gets at least one more of those points than you do. So, if you can win by offering a (long) n-chain to the canonical guru, you'll have to be at least $n - 1$ points ahead when you play that loony move.

Mathematical Proofs Using "The Man in the Middle"

There is a interesting technique which can be used to prove all of the assertions made so far in this chapter, as well as certain other theorems. This technique is called the "man-in-the-middle".

A brash mathematical WhizKid, W.K., who has barely learned the rules of chess, boasts that he is ready to beat some famous grandmaster such as Judit Polgar or Gary Kasparov. W.K. asks only that the grandmaster grants him a modest time advantage on the clocks: 2.5 hours for W.K. versus 1 hour for the grandmaster. An incredulous patsy wagers a large sum of money that the whizkid will surely lose to both grandmasters. The patsy becomes even more confident when he learns that the whizkid will play both of them simultaneously!

The games begin as shown in the following table[†]. The first column indicates the time at which each move is made.

	Board 1		Board 2	
	Gary Kasparov (white)	W.K. (black)	W.K. (white)	Judit Polgar (black)
time				
00:00	1. e4			
00:01			1. e4	
00:02				c5
00:03		c5		
00:04	2. Nf3			
00:05			2. Nf3	
00:06				d6
00:07		d6		
00:08	3. d4			
00:09			3. d4	
...				

[†] This chess opening, without any man in the middle, was actually played between Gary Kasparov and Judit Polgar in the Corus 2000 Tournament at Wijkanzee, Netherlands.

The patsy soon realizes that he will lose his bet, because W.K. has cleverly positioned himself as "the man in the middle" of what amounts to a single game! The grandmasters must compete with each other, removing any possibility that BOTH will win.

When applied to a canonical Dots-and-Boxes assertion, the "man in the middle" argument is used to refute the assertion that a non-canonical move can win in some position in which the corresponding canonical move loses. The identical position is setup on Boards 1 and 2. Guru "Kasparov" begins with his presumed noncanonical move on board 1. The man in the middle then plays his corresponding canonical move on board 2. The man in the middle then copies the response of guru "Pulgar" onto the other game against guru "Kasparov". This works so long as the gurus play in regions not directly adjacent to the non-canonical opening move and its canonical facsimile. For each theorem, careful checking may be needed to ensure that when a guru makes such a move, the man in the middle is able to respond in a way that removes the small discrepancies between the boards, and that he does so in such a way that either leaves the net scores the same or, in some cases, might yield a net advantage for the man in the middle.

Our most assiduous reader will no doubt be able to verify all of the relevant details. Other readers may prefer to accept these assertions without proofs.

Some Elementary Endgame Positions with Few Joints

Figure 33 shows several endgame positions which commonly occur as summands in larger games: An independent chain of length 5, an independent loop of length 8, a "dipper" with a handle of length 4 and a cup of length 4, and "earmuffs" consisting of two (dependent) loops of length 4 connected by a headband of length 6.

A "joint" is a node of valence ≥ 3. The loop and the independent chain contain no joints; the dipper contains one; the earmuffs, two. Although the dipper and the earmuffs each contains a chain of length n, only the chain which runs from ground to ground is "independent".

We assert that

> A canonical player will never play an independent chain if a shorter independent chain is available
>
> and
>
> A canonical player will never play a loop if a shorter loop is available.

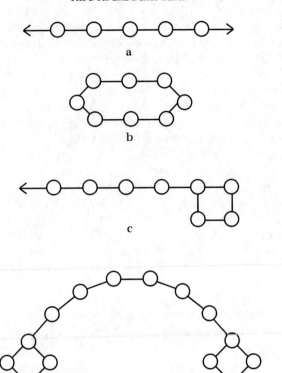

Figure 33. (a) a long chain, (b) a loop, (c) a Dipper, and (d) Earmuffs.

In particular, a canonical player will never play an independent long chain (of length ≥ 3) if any short chain (of length ≤ 2) is available.

The man-in-the-middle argument can be used to verify that these restrictions will never prevent our canonical guru from doing as well as is possible against any (possibly non-canonical) opponent.

Canonical Comparison of Loops and Chains

Assertion:

> If a loony endgame contains a loop of length k and an independent chain of length n, and if $n \geq k$, then a canonical player should always prefer playing on the loop to playing on the chain, except possibly when $k = n = 4$.

W.l.o.g., we may assume that k is the length of the smallest loop and that n is the length of the smallest chain. The proof, which uses both the man-in-the-middle argument as well as several calculations with values, appears in [8 and 9].

An Overview of a Typical Large Dots-and-Boxes Game

Certain milestones will occur in the course of any Dots-and-Boxes game. When the board is large, and the game is played between two canonical gurus, I would not be surprised to see these milestones occur in the following order:

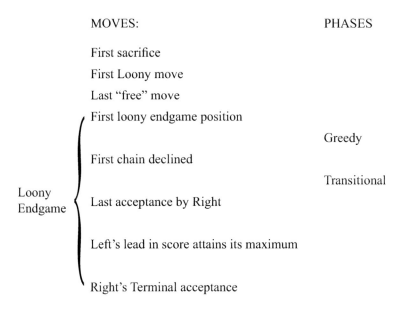

Experts often begin sacrificing much earlier than beginners. Typically, after one player has made several such sacrifices, the other player will play a loony move, even though several "free" moves (which cost no points) may remain. Usually this early loony move is a 3-chain or a 4-chain near the middle of the board. If it is not sacrificed early, the opponent will keep control while extending this chain to a much longer length. After the early sacrifice, the areas at both ends of the chain come into play. Whether this early loony sacrifice should be accepted or not can be a challenging question, especially when the score is destined to be very close.

The rest of this chapter is devoted to the plays which occur after the last move whose incentive was not loony.

Values of Loony Endgames

We say that an endgame becomes "loony" when only loony moves remain. This is distinct from a "loony position", which necessarily occurs after the next move. The nimstring value of a loony endgame is 0. Nearly all Dots-and-Boxes games become loony after sufficiently many turns. The play of the loony endgame typically involves a large fraction of the total number of points scored in the entire game, even though it usually lasts a relatively smaller number of turns.

If we count only the net score of the remaining (loony) position, it is clear that this cannot favor the player who needs to make the next (loony) move, because his opponent will soon have a choice, and often some additional points as well. So we define the Strings-and-Coins *"value"* of a loony endgame position G, denoted by $V(G)$, as the net advantage in score which that position provides to the second player. Suppose that first player moves on a chain of length c (which need not be independent). Suppose further that after all of this chain has been taken, then the graph which remains is called G'. Then

$$V(G) = c - 2 + |V(G') - 2|$$

because the second player will take $c - 2$ boxes immediately, and then choose between declining the last two boxes in order to score $V(G') - 2$ additional points, or accepting the last two boxes in order to score $2 - V(G')$ additional points.

Similarly, if first player moves on a loop of length k, leaving a position G', then

$$V(G) = k - 4 + |V(G') - 4|$$

In general, if first player has a choice of several long chains and loops on which to play his loony move, then the value of G is the minimum among all such expressions.

We notice in particular that if a loony endgame has value 0, then the first move must necessarily be on a 4-loop. After the 4-loop is taken, the new graph, G', must have Strings-and-Coins value 4.

Fully Controlled Values

Loony Dots-and-Boxes endgames often reach a stage in which the same player remains in control for a substantial number of consecutive moves. Either Arthur or Beth might become this controlling player. This player might be viewed as the dictator, the boss, or master; his opponent as the rebel, the worker, or the slave. We prefer the simpler terms *Right* (for the player in control) versus *Left*. At each turn, Left must make a loony move. Right then takes all-but-two of the offered chain or all-but-four of the offered loop, and remains in control, leaving Left with another loony endgame.

Let us now introduce the simplifying assumption that Right is (artificially) compelled to remain in control, even to the point of forcing Left to take the very last box of the game. The net score (for Right) under this assumption, is called the "fully controlled value".

For most positions that occur on graphs which are not unreasonably large, fully controlled values are reasonably straightforward to compute. As in prior chapters, the number of strings attached to any coin in the graph is called its *valence*. Coins with valence ≥ 3 are called *joints*. Joints and the ground are called *stops*. Then every chain runs from a stop to another (possibly the same) stop.

Suppose the graph has c coins and j joints with total valence v, counting grounded ends as having valence 1 each. Let's further suppose that in the course of the game, Left makes m moves on isolated cycles (loops) and n moves on chains. So Left's total score is

$$4m + 2n$$

and Right's total score is

$$c - 4m - 2n$$

A move on an isolated cycle doesn't change the valence, but a move on a chain decreases the valence by 1 at each end, except that whenever the value of a joint changes from 3 to 2 that joint disappears. That happens just once for each joint, so

$$v = 2n + 2j.$$

and so Left's total score will be

$$4m + v - 2j$$

Since v and j are fixed, Left will scheme to make as many moves on isolated cycles as possible. If m is the maximum number of such cycles, the fully controlled value is

$$c + 4j - 2v - 8m$$

The fully controlled value of a position which is the sum of several regions is easily seen to be the sum of the fully controlled values of the regions. The fully controlled value of a (grounded) chain is its length $- 4$. The fully controlled value of a loop is its length − 8. The fully controlled values of the dipper and the earmuffs in Figure 33 can be determined by viewing each position as a sum of a chain and loops. From the perspective of fully controlled values, these simple positions decompose just as they did in Chapter 6.

"Very Long" Defined

A very long chain or a very long loop is one whose fully controlled value is non-negative. That means a chain of length at least 4, or a loop of length at least 8. Chains of length 3 and loops of lengths at least 4 but less than 8 are called "not-very long".

Right's Terminal Bonus

Even if Right is not artificially compelled to retain control, he will do so voluntarily if enough sufficiently long chains and loops remain to be played. Left will generally choose to leave these until near the end. However, eventually the last very long string will be played, and if it is to Right's advantage to accept all of it, he will do so and gain an improvement over the fully controlled score. This gain is called Right's terminal bonus. Here are its values for several endings:

Terminal Position	Bonus
Sum of independent chains of length 3	4
Independent Chain of length ≥ 4	4
Long Loop + 3-chain	6
Long Loop	8

In order to minimize Right's terminal bonus, if possible, Left should leave a terminal position consisting of a longest independent chain. If there are no very long independent chains, this might be a 3-chain.

Controlled Values

The "controlled value" of a loony endgame G is its fully controlled value plus Rights' terminal bonus. We denote this be $CV(G)$. Since Right, if he so chooses, can elect to retain control until he can take his terminal bonus, it is clear that

$$V(G) \geq CV(G)$$

The controlled value can often be computed much more quickly than the actual value. It also turns out that, under very common conditions we shall discuss later, the two are known to be equal. Since the controlled value is so useful, it is very convenient to express it in the following formula, which combines the formulas for the fully controlled value and Right's terminal bonus into a single expression:

$$\text{If } p > 1/4, \text{ then } CV(G) = 8 + c + 4j - 2v - 8p$$

where p is the maximum adjusted number of node-disjoint loops, obtained by counting each loop as

      1 if it excludes the ground
      ½ if it includes the ground and at least four other nodes
      ¼ if it includes the ground and three other nodes

Since the ground is a single uncapturable node, and independent chains are viewed as loops through the ground, at most one chain can be included in the count of p. So $p = 1/4$ only in the degenerate case in which G consists entirely of a sum of independent chains of length 3.

When Strings-and-Coins Value Equals Controlled Value

If G contains many not-very long strings and loops, then its controlled value can be negative. In such a case, it is obvious that $V(G) > CV(G)$. The best play may offer some not-very-long chain or loop, a gift which is accepted. Good players may continue to exchange such gifts for several moves, but eventually one player will elect to take control and keep it until he can claim his terminal bonus. This player assumes the role of Right.

After each loony move by Left, Right can decide whether to accept or decline. Left's goal is to order her moves in such a way that Right cannot gain by any acceptance prior to the last or next-to-last loony move. If Left is unable to find such an order, then $V > CV$.

Left's best strategy during the controlled phase of the loony endgame is often to pick an appropriate future position, H, and make plays which increase Left's score while moving towards it. Ideally, all chains and loops in H should be very long, thus ensuring that once H is reached, playing out the rest of the game will be very straightforward. It is also desirable that H have a high enough value to ensure that Right is not tempted to abandon control before the position reaches H. Clearly, if the number of coins in G is c, then $CV(H) \geq c/2$ is sufficient. Actually there is a sense in which the much weaker condition $CV(H) \geq 10$ is now known to be sufficient, even if c is very large [8]. To be specific, suppose that no node of G has valence > 4, and suppose that L is a maximal set of node-disjoint loops in G. Then suppose that H contains none of the not-very-long loops in L, nor any of the chains in G which touch any of those not-very-long loops in L. Suppose further that H contains no not-very long chains in H. If $CV(H) \geq 10$ and $CV(G) \geq 10$, then $V(G) = CV(G)$.

To see that no general result can be significantly stronger than this, we note that a chain of length 9 plus a pair of earmuffs (both 4-loops) connected by a headband of length 11 has controlled value 8 but actual value 10. However, in the common but simple case in which G consists of a sum of only independent chains and loops of even lengths, it has been shown that if $CV(G) \geq 2$, then $V(G) = CV(G)$. In this case one can take H to be the sum of those components of G which are very long.

For Big Games, even Finding the Controlled Value Can Be Hard

Our formula for the controlled value includes a term p, which is the maximum adjusted number of node-disjoint loops in G. For graphs of modest size, this is easy to compute because the number of joints, j, is typically quite small. Problem 11.24, which is shown on the cover of this book is a relatively complicated 9×9 Dots-and-Boxes game, yet it soon leads to a loony endgame position with only 6 joints. The number of joints in typical Dots-and-Boxes endgames seems to grow quite slowly with the size of the board.

Yet when the number of joints becomes large, the computation of p may become intractable. Even in the case of Swedish positions, which contain no ground node and therefore no adjustments, a serious difficulty remains. The problem of finding a maximal set of node-disjoint loops on an arbitrary (possibly nonplanar) large graph is NP hard. That is a technical term well beyond the scope of child's play. It means that no one in the world knows a good algorithm which provides a general solution to this problem without running for a very long time, even on a very fast computer.

Chapter 11
Dots-and-Boxes Problems with Close Scores

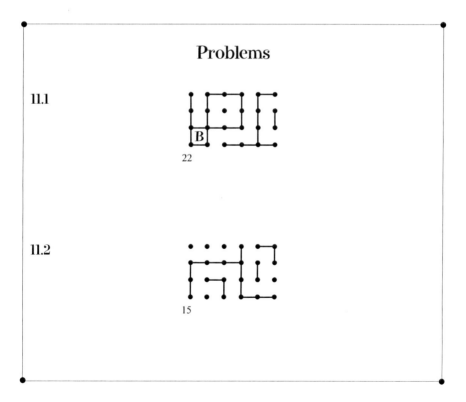

Problems

11.1

22

11.2

15

Answers

11.1

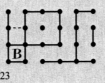

B should take this chain, offer the loop, and win 9–6, because the last 12 boxes will be split 6–6.

11.2

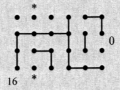

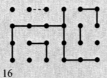

B cannot get control, but she can win 8–7.

Problems

11.3

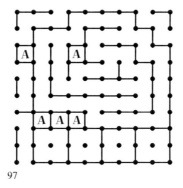

97

In this problem, determine the final score if both players play well. Explain why. Is there any move which does *not* win? Where and why?

11.4

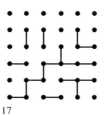

17

Answers

11.3

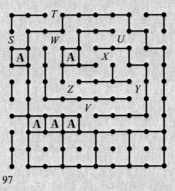

A can win 41–40 by playing *UX, VY, WZ*, the three 3-chains on the boundary, the four 4-loops along the bottom. He should save *ST* until the very end. Premature play on *UV, UW, XY, YZ, XZ*, or *ST*, loses.

11.4

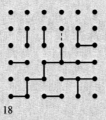

Problems

11.5

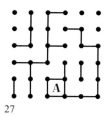

27

11.6

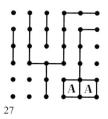

27

Answers

11.5

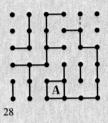

28

Now there are 2 free moves and 3 chains of length 4. *A* will lead by 3 points when only loony moves remain. Giving both chains adjoining the loop allows *A* to win 13–12.

11.6

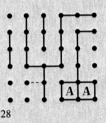

28

The dashed move sacrifices a point immediately, but it takes control and allows *B* to win the game, 13–12.

Problems

11.7

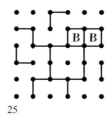

25

11.8

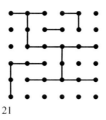

21

Answers

11.7

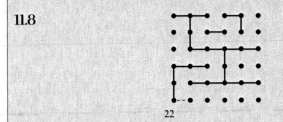
26

To keep control, *A* must next sacrifice 2 in east, but then *B* plays a loony move in north!

11.8

22

A single move in either the north or the east can create either a chain or a loop. But if either player plays such a move, his opponent will play the other region, to his advantage. So each player should play elsewhere as long as feasible.

Formally, the nimbers are as follows:

North (east)	(8 boxes)	2
East	(4 boxes)	2
Northwest		0
South		1

The nimber is 1, so *B* can retain control by playing a move of incentive 1, such as the move shown.

This move also extends the long chain, which the controller will eventually take.

There are many available moves with incentive *. They include a pair of moves which sacrifice the single box in the northeast corner, or another pair of moves which create a dangling 2-chain there. Later, there may also be some moves that sacrifice one or two boxes in the south or northwest. However, eventually all such moves will be gone. *A* will then be forced to play a move with incentive *2 or *3, in the north or east, and *B* will then play the other to retain control and win the loony endgame, and the game.

Problems

11.9

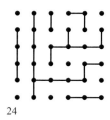

24

11.10

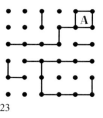

23

Answers

11.9

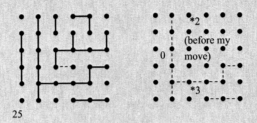

25

It is necessary to make a move of incentive *, *not* a move of incentive *2 or *3. The marked move (or any other move from the dot at the marked move's east endpoint) is preferred, as it evades a threatened loop to ensure a higher winning score.

11.10

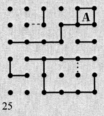

25

Play dashed move in the northwest to take control. If *B* then plays the dotted move to stop the 6-loop, *A* extends the chain into the southwest corner. *A* may then need to give 2 in the southeast, rather than 0 or 1 in the northwest, in order to ensure winning the game.

Problems

11.11

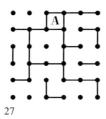

27

11.12

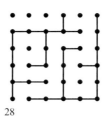

28

Answers

11.11

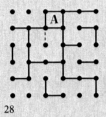

28

(1) The only winning move is to sacrifice the 4-loop immediately. *B* can now choose to either play "*X*" or "*Y*" for the rest of the game.

(2) *X* takes 4 boxes and then sacrifices the southern 3-chain before it grows any longer. *Y* will decline. There remain 3 free moves (on each in east, NW and SW), 3 short chains of length 2, and 2 big chains of lengths 3 and 8. *Y* keeps control and gets the 8-chain, but *X* and *Y* end with 12 points each. So *A* wins the game, 13–12, no matter whether *B* chooses to play "*X*" or "*Y*."

32

(3) If instead of sacrificing the 4-loop, *A* begins by offering the southern 3-chain, *B* could accept this sacrifice and then offer the 4-loop and win.

11.12

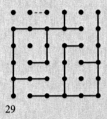

29

A's dashed move threatens a second chain, so *B* must sacrifice two boxes in the north to retain control. *A* then plays a loony move to sacrifice the center square. *B* can either take this 3-chain or decline it, but either way *A* eventually wins 13–12.

Problems

11.13

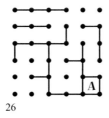

26

11.14

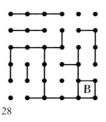

28

Answers

11.13

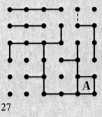
27

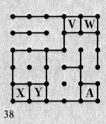
38

After boxes A, V, W, X, and Y are taken, the last 20 points will split 10–10. So to win, *B* must get at least three of the four points at V, W, X, and Y. The move shown splits V and W so that *B* can eventually get one of them plus X and Y to win the game 13–12.

11.14

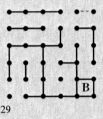
29

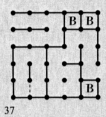
37

Note the similarities to the prior problem. After several "free" moves, *A* will give 2 boxes in the northeast corner to *B*, who will take them and offer the southwest 6-loop. After taking 2 of these 6 boxes, *A* loses 13–12 as in the prior problem.

Problems

11.15

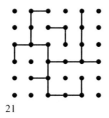

21

11.16

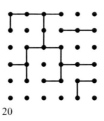

20

Answers

11.15

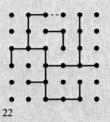

22

As in the prior two somewhat similar problems, the normal loony endgame will divide the last 16 boxes evenly.

If a 3-chain is created in the northeast, then B can win by assuming control until the loony endgame begins. If no 3-chain appears there, then B can win by collecting a majority of the 9 points on short chains. So the northeast is irrelevant. B's abnormal move ensures victory.

11.16

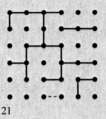

21

There will be one chain in the northeast. With two loops, the loony endgame won't be worth enough to win unless the southeastern chain is fairly long. The dashed move threatens to make a 5-chain here, while taking control and winning. The cost to B of taking control is too much.

Problems

11.17

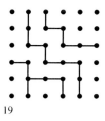

19

11.18

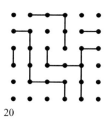

20

Answers

11.17

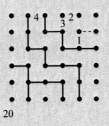

20

The dashed move forces *A* to play 1. *B* takes two boxes and plays two, forcing three. *B* takes two more boxes and plays the loony move at four. *A* takes three boxes, but then, whether *A* elects to double-deal or not, the last 18 boxes divide evenly.

11.18

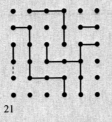

21

Take control. If *B* next plays either dashed move shown below, respond by playing the other, and then matching southeast with northwest.

25

Or, if *B* sacrifices the center, play the dotted move and then match southeast with northeast.

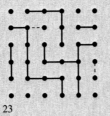

23

Problems

11.19

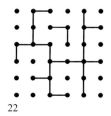

22

(Advanced)

Answers

11.19 In this problem with A to move, A clearly has control. The north is *2, and every other region is 0 or *1. We might normally assume A should make a loop in the north in order to ensure 0 chains.

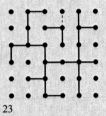

23

But this is an abnormal situation in which A wins *only* by forcing an odd number of chains! That's because the loony endgame will be two 4-loops and either an 8-loop or an 8-chain, and either of these final configurations of 16 boxes will be evenly divided, 8 to 8. Under normal play, whoever moves first on this loony endgame will have just completed the last (and decisive) short chain, and the move shown is the only one on the board which ensures that A can become this victorious player.

In general, when the loony end-position is sufficiently close, the game can be analogous to misère Nim [1]. When only one Nim-heap contains 2 or more chips, the player to move can determine who gets the last chip. In Dots-and-Boxes, this "last chip" corresponds to the last non-loony move, which gives one's opponent the last short chain. Under "normal" conditions, this is a desirable outcome, because the opponent is then forced to give back long loops and chains. However, in the abnormal case, when the long loops and chains combine into a very close endgame, it may be advantageous to get the last short chain rather than the first long one.

Problems

11.20

31

Answers

11.20

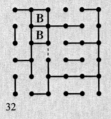

32

There are 4 chains, so *A* has control. Although there are two "free" moves, *B* must instead play a loony move, giving away a 3-chain.

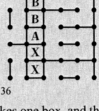

36

A takes one box, and then faces a Hobbsian choice: to play "*X*" or "*Y*." It matters not, because *X* and *Y* will each get 11 of the last 22 points, so that *B* wins 13–12. After taking two, *X* takes the only free move.

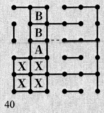

40

Y gives *X* two in the southwest. *X* takes and then gives the northeast. This isolates the 6-loop.

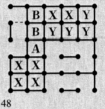

48

Y takes all but the last two. *X* takes these and gives back one 3-chain.

52

Y takes one but declines the last two. *X* takes, and then offers the 6-loop rather than the remaining 3-chain!

61

Declining the loop would cost *Y* four of its boxes, so *Y* takes the entire loop but must then allow *X* to take all of the last 3-chain.

Problems

11.21

22

Answers

11.21 This problem is partitioned into four regions with the following nimbers:

Northwest	(8 boxes)	0
Northeast	(6 boxes)	2
East	(6 boxes)	2
South	(5 boxes)	0

To see that south is 0, it is sufficient to verify that either the first or the second player can ensure that no long chains are made there.

The problem is very challenging for A, because B has control.

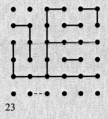

23

After this move, the south has nimber 4. Here are the nimbers of some potential southern positions: Those that are not immediate followers are in parentheses.

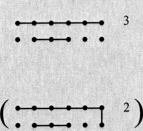

Answers

(11.21 continued)

 2

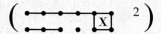 2)

•———•———•———•———• 1(*d*)
•   •———•   •———•   •

•———•———•———•———• 0(*v*)
•   X   •   •   •

In order to retain control, *B* must play a move with incentive *4. There are no such moves in any other region, so *B* must sacrifice the box on which *A* just played. *A* accepts and continues like this:

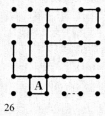

26

Answers

(11.21 continued)

The total south now has nimber 3 (*1 in the southwest + *2 in the southeast). To retain control, B must play a move with incentive *3. There are no such moves in the northwest; all moves there have incentive *. Here is a move of incentive *3 in each of the eastern regions:

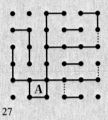

27

After B plays such a move, A plays another. Specifically, he plays in the northernmore of the two remaining eastern regions. B can then retain control only by playing the third of these moves. A collects two more boxes in the south and plays in the northwest as shown:

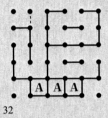

32

There are then five turns of incentive *, which we view as "daisy" moves, as in "She-Loves-Me; She-Loves-Me-Not" [1]. A acquires 2 more points and then leads by a score of 5–1. The loony endgame consists of one 7-chain and two 6-loops. B retains control, but A collects 4 points on each loop and wins the game 13–12!

Nor can B do any better by conceding control. If A gets control, he also gets a second chain, which ensures him a bigger total score.

Problems

11.22

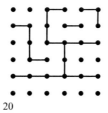

20

Answers

11.22 Resign?! At first glance, this problem looks less challenging for *A* than the previous one, because although *B* still has control, the controlling grip in the east appears weaker.

However, we believe that *A* has no winning move in this problem position. His attack begins as before, but *B* now plays a different response:

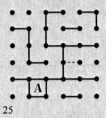
25

This move has incentive *3, so *B* retains control. *A* takes two boxes, plays in the northeast, and *B* then sacrifices two more boxes in the south. After taking them, *A* nibbles at the controller's weakness in the east by playing the dotted move.

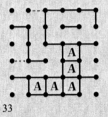
33

B responds with the dashed move in the northwest. Then, both players take a free move in the north, one box in the south, and a pair of boxes in east or west. *A* then has 8 points, and can score only 4 more in the loony endgame, so *B* prevails.

Problems

11.23

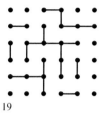

19

11.24

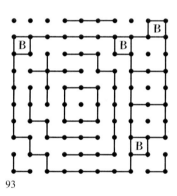

93

Answers

11.23

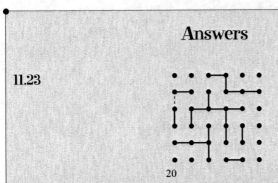

From the nimstring perspective, each of the two long chains in this problem can be split, separating the southeast from the (north) west. The chain of length 2 in the southeast will either be sacrificed or joined onto one (but not both!) of the long chains. So the southeast nimber is 0 or 1. The (north) west nimber is at least 2, so this contains a canonical move which wins at nimstring.

The move shown forces one long chain, and eventually victory for *B*, although early loony moves by *A* yield a close score. Compare with Problem 5.14.

11.24

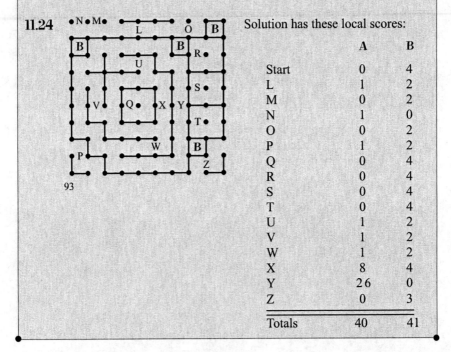

Solution has these local scores:

	A	B
Start	0	4
L	1	2
M	0	2
N	1	0
O	0	2
P	1	2
Q	0	4
R	0	4
S	0	4
T	0	4
U	1	2
V	1	2
W	1	2
X	8	4
Y	26	0
Z	0	3
Totals	40	41

Chapter 12
Unsolved Problems

Unsolved Problems

12.1

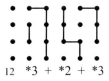

12   *3 + *2 + *3

Although there are several moves that keep control, none of them lead to a winning endgame. Does any other move win?

Unsolved Problems

12.2

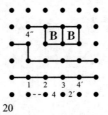

20

After the dashed move and turns 1–4 as shown above, A can choose to play either X or Y in the following position, with X to move.

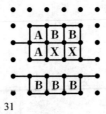

31

With best play, X can now find sufficient threats to turn the north into five X boxes plus a 2-chain and a 1-chain, but the last 16 boxes are evenly divided between X and Y. So A does best to play X, but B wins by at least one point.

So perhaps A should not sacrifice at move 3 (?).

12.3

Does it matter whether X is A or B?

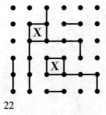

22

Unsolved Problems

12.4

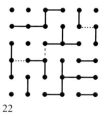

22

Dashed move once looked worthy, with both dotted moves likely to be played soon thereafter.

12.5

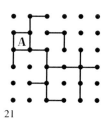

21

Closely related to other problems.

12.6

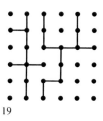

19

Unsolved Problems

12.7

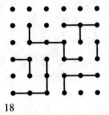

18

12.8

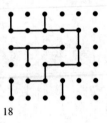

18

12.9

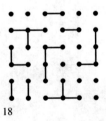

18

Unsolved Problems

12.10

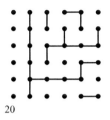

20

B has control because west = 0, south = *2 = northeast.

12.11

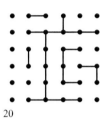

20

12.12

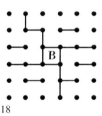

18

Unsolved Problems

12.13

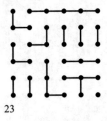

23

Free moves left = 11 vertical + 2 horizontal.
Many moves remain but still connected.

12.14

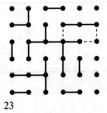

23

Suggested approach: Consider each of the dotted moves individually. Since loony chains split, any of these moves cuts the nimstring game into two big pieces of about 8 active nodes each. Each of these positions which does not achieve control has (several?) followers which do. The "follower" move cannot attain control from the initial position, because it can be answered by one of the three dotted moves shown.

This method should facilitate a nimstring analysis, which might provide the basis for a Dots-and-Boxes analysis.

Unsolved Problems

12.15

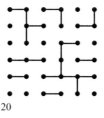

20

12.16

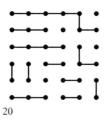

20

12.17

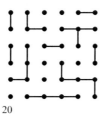

20

Unsolved Problems

12.18

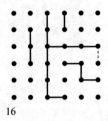

16

Dashed move once looked promising.

12.19

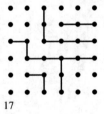

17

Bibliography

[1] Elwyn Berlekamp, John Conway, and Richard Guy,[1982] *Winning Ways:For Your Mathematical Plays*, (Volumes I and II), Academic Press, London and New York.
   This opus consists of four parts, first published as two volumes with two parts per volume. The Dots-and-Boxes game is the focus of Chapter 16, which appears in Part 3. A K Peters, Ltd. will soon publish a new edition in four volumes, one part per volume. Although there is substantial overlap between this book and that chapter, each also contains a significant amount of material not found in the other. Topics in *Winning Ways* which are not in this book include an analysis of the 4-box game, when to keep control when there are few long loops and chains, strings-and-coins values of vines, nimbers for nimstring arrays, and 12 more problems.*Winning Ways* also includes extensive background material for the serious student. See [3] below.

[2] Elwyn Berlekamp and David Wolfe, [1994] *Mathematical Go: Chilling Gets the Last Point*, A K Peters Ltd., Natick, MA.
   This book extends some results of *Winning Ways* [1] and applies them to the world's most-played classical board game. The incredible lack of overlap between Go and Dots-and-Boxes provides a compelling illustration of the breadth and power of combinatorial game theory.

[3] John Conway, [1976] *On Numbers and Games*, Academic Press, London and New York.
   The writing of this book began several years after *Winning Ways* began, and was finished many years earlier. It is a surprisingly accessible mathematical tour de force. It includes a rigorous treatment of "surreal numbers," superseding the classical constructions of Dedekind and Cantor.

Bibliography

[4] Richard K. Guy, [1989] *Fair Game: How to Play Impartial Combinatorial Games*, COMAP Mathematical Exploration Series, COMAP, Inc., 60 Lowell Street, Arlington MA, 1989. ISBN 0-912843-16-0.

This book is a spinoff of Chapter 4 of *Winning Ways*, which, in turn, was heavily based on Guy's earlier work. It provides superb introductory exposition of impartial games, including many topics related to Dots-and-Boxes, ranging from the classical daisy-petal counting She-Loves-Me, She-Loves-Me Not; through normal and misere nim, through Kayles and Dawson, to subtle issues of more general periodicity.

[5] Richard K. Guy, [1991] *Combinatorial Games* (Proc. Short Course on Combin. Games, Columbus OH, Aug 1990, AMS Proc. Symp. Appl. Math. 43

This includes a superb introduction to Dots-and-Boxes by Richard Nowakowski.

[6] Sam Loyd, [1914] "Cyclopedia of 500 Puzzles, Tricks, and Conundrums with Answers," reprinted by Corwin Books, NY, 1976.

Page 104 shows a Dots & Boxes problem on the level of Chapter 1 of this booklet.

[7] Richard Nowakowski, ed., [1996] *Games of No Chance*, Cambridge University Press, Mathematical Sciences Research Institute Publications, Cambridge, UK, and New York.

This rich collection of papers about combinatorial games includes Julian West's commentary on the final match of a 5 × 5 Dots-and-Boxes tournament, played between Daniel Allcock and Martin Weber.

[8] Richard Nowakowski, ed., [2001] *Proceedings of the July 2000 Workshop on Combinatorial Games*, Mathematical Sciences Research Institute Publications, Cambridge University Press, Cambridge, UK, and New York.

This sequel to [7] includes the latest results of Elwyn Berlekamp and Katherine Scott on loony Dots-and-Boxes endgames.

[9] Katherine Scott, [2000] "Loony Endgames in Dots & Boxes," MS thesis, University of California at Berkeley.

Index

ℂ = "loony", 61
* prefix, 45
*, "star", 75
*1, 75
* ≥ 5, 57
6-box sacrifice, 20
6-loop, 100, 23
9-box game, 9
1 × 5 semi-Icelandic, 110, 111
2 × 2 Icelandic game, 46, 49
3 × 3 problems, 15
3 × 5 problems, 17, 19, 53, 87, 117
5 × 12 problems, 73
5 × 8 problems, 71
6 × 10 problems, 73
9 × 9 problems, 89, 116

A

A, the first player, 7
abnormal endgame, 106
adjusted number of node-disjoint loops, 85
advanced chain counting, 23, 27
advanced Nimstring problems, 69
all-but-four of each long loop, 77
Allcock, Daniel, 126
Amy, 3
arrays, 59
Arthur, 3, 9

B

B, the second player, 7
Babar, 3
Baldwin Wallace College, xi
Bell Telephone Laboratries, xi
Berkeley, University of California at, x, xi
Beth, 3, 9
BIG, 51
big nimber, 50, 76
binary, 49
Bouton, Charles L., xi, xii, 49

C

Cantor, 125
Calgary, University of, xi
California, University at Berkeley, x, xi
canonical, 76, 77
    comparison of loops and chains, 80
    nimstring, 116
    player, 79
    strings-and-coins, 76
capturable coin, 41
cast off long loops and chains!, 63

Index

ceding control to win, 106
chain, 8
  = "long chain", 1
  counting problems, 13
  infer parity of number, 51
  of twopins-vine, 65
  side, 16
chess, 78
choice, canonical, 77
chopping and changing, 61
coins, 11, 41
complimentary moves, 42
computer, 67
computer game-playing, x
control, 5, 6, 10
controlled value, 84
  hard, 86
  equals strings-and-coins value, 85
  of loony endgame, 84
controlling player, 82
conventions for problems, 1
Conway, John Horton, xi, xx11, 125
cost of control, excessive, 102
count of moves played, 1
cover problem, 115, 116
crosses, 64
cycle = loop, 40

D

D4, D9, D11, *See "Dawson's, vines"*
daisy moves, 112, 126
Dawson's
  Kayles, 67, 126
  recursion, 65
  values, 66
  vines 65
  D11, 74;
  D4, 72;
  D9, 74;
decline or take, 41, 44, 77
declining a handout, 6
declining a long loop, 23
decomposing vines, 66, 67
Dedekind, 125

dipper, 79, 80
dots and doublecrosses, parity, 7
doublecross, 6, 7, 42
  none yet convention, 1
double-dealing, 5, 6
double-jointed vines, 64
dual of dots-and-boxes, 11

E

earmuffs, 79, 80, 86
elementary nimber problems, 53
elephant king, *See "Babar"*
endgames with few joints, 79
enough rope principle, 9
equivalent corner moves, 12
even endgames
  16-point, 102
  18-point, 104
  20-point, 100
exceptional values, 66
exclusive-or = XOR, 49, 50

F

Fair Game, 45, 125
first
  chain declined, 81
  loony endgame position, 81
  loony move, 81
  sacrifice, 81
force one chain, 52
force number of chains, 25
free coin, 42
free your fetters!, 63
fully controlled values, 82, 83

G

Gardner, Martin, x
Go = Asian board game, 125
graph, Dawson's vine, 67
graph, strings-and-coins, 11
greedy phase of loony endgame, 81
Grossman, J.P., x
ground, 25

Index

Grundy, Patrick Michael, xi, 125
gurus, omniscient, 24
Guy, Richard, x, xi, 45, 125, 126

H

H, future position, 85
half-hearted handout, 8, 16, 24, 42, 43
   NOT canonical, 76
hard nimstring problems, 47
hard-hearted handout, 8, 76
harmless mutation theorem, xi, 60, 61
Harvard, 49
headband, 79, 86
Hogan, Apollo, xii
horizontal corner moves, 12
How long is long?, 8

I

Icelandic game, 10
   1×5 semi, 110, 111
   2×2, 46, 49
impartial games, 45, 67, 126
incentive in nim, 75
   *, 94, 96, 112,
   *2, 96
   *4, 111
independent chain, 79
infer chain-count parity, 51
initials in boxes, 3
irrelevant region, 102

J

jagged corner problem, 57
joints, 24, 65, 79, 83

K

Kasparov, Gary, 78
Kayles, 66, 67, 126
   vines, 65
      with 5 tendrils, 70
Kelly, J. L., Jr., xii
killing mutation, 60

L

L = maximal set of node-disjoint loops, 86
last acceptance by right, 81
last free move, 81
Lauria, Tony, xi
Left, as controllee, 82
Left's maximum lead in score, 81
Little Women, 3
local properties, 42
long chains, 6, 8, 80
   = branches − nodes, 25
   = branches − nodes (mod 2), 51
   abbreviated as "chains", 1
   casting off, 63
   notation, 60
   rule, 7, 9
   sameness of, 59
   snap, 61, 65
   take all but two, 40
long cycle, take all but four, 40
long is long, how, 8
long is long, why, 23
long loops, casting off, 63
loony = $\mathbb{C}$, 61
   endgame, phases of, 81
   move, 24
   move is a losing move, 77
   position, 82
   value, 44, 49
loop, 80
lose your shackles!, 63
Loyd, Sam, 126

M

man in the middle, 78, 80, 81
Mang, Freddy, xii, 52
maximize our score, 70
mex function, 44
milestones of a game, 81
minimal excludant, mex, 44
misere Nim, 106, 126
MIT, x

Index

move count, 1
mutations, 59, 60

N

nim, x, 49, 126
nimbers
  & chain-counts, 50
  followers, 75
  problems, 53
  values, 39
nimstring, x, xi, 10, 59
  equivalencies, 61, 63
  graphs, 39
  incentives, 75
  program, on-line, 52
  value, 44, 49, 57
non-canonical, 76
nonplanar large graph is NP hard, 86
normal play rule, 39
noteworthy nimstring nimbers, 64
not-very long, 84
Nowakowski, Richard, 126
NP hard, xi, 86
numbers, surreal, 125

O

OMAR, 79
omniscient gurus, 24
on-line nimstring program, 52
Our Most Assiduous Reader, 79
overview of large game, 81

P

p, the maximum adjusted number of
  node-disjoint loops, 85
parity
  of branches, 51
  of dots and doublecrosses, 7
  of long chains, 51
  of nodes, 51
partizan, xi
path, 61
periods of D and K, 66, 126
phases, 81
Polgar, Judit, 78

problems
  3×3, 15
  3×5, 17, 19, 53, 87, 117
  5×12, 73
  5×8, 71
  6×10, 73
  9×9, 89, 116
  hard nimstring, 47
  unsolved, 117
  with chain counts, 13, 27
  with close scores, 87
  with nimbers, 53

R

recursion, Dawson's, 65, 66, 45, 44, 51
recursive computations, 51
resign!?, 114
Riggle, Timothy, xi
Right, as controller, 82
  terminal acceptance, 81
  terminal bonus, 84
rules for loony values, 49

S

sacrifice, 16, 98
Schleimer, Saul, xii
Scientific American, x
scissors, 11
score, dependence on loop count, 83
Scott, Katherine, xii, 126
she-loves-me, she-loves-me-not, 112, 126
short chain, 9, 23, 80
short loop, 23
side chain, 16
single-jointed vines, 64
Smith, Cedric, xi, xx, 10
snap long chains, 63, 65
special notation for long chains, 60
spokes, 9
Sprague-Grundy theory, x
Sprague, R. P., xi
stems of vines, 64, 65
stops, 83
strategy stealing, 24, 43

stretch your limbs, 63
strings-and-coins, 11, 39, 82
    value equals controlled value, 85
sums of nimstring positions, 49
surreal numbers, 125
Swedish positions, 10, 86

T

take all-but-two, 77
take or decline, 41, 44, 77
take or win, 44
tendril, 64, 65
The Tech (MIT newspaper), x
three chains in 16 moves, 38
three or more's a crowd, 59
tic-tac-toe, xii
to take or not to take
    in nimstring, 41
    in strings-and-coins, 77
transitional phase of loony endgame, 81
triangular boards, 10
turns in the game counted, 7
twopins vine, 65, 67, 72

U

unjointed vine, 64
unresolved chain count, 26, 51
unsolved problems, 117

V

valence, 24, 83

value, loony nimstring, 82
value, loony strings-and-coins, 82
vertical corner moves, 12
very close scores, 75
very long chain, 84
vines, xi, 59, 64, 65
    Dawson's, 65
    decomposing, 67
    double-jointed, 64
    Kayles, 65
    single-jointed, 64
    stems, 64, 65
    tendrils, 64, 65
    twopins, 67
    unjointed, 64

W

Weber. Martin, 126
West, Julian, 126
WhizKid, 78
Who wins nimstring?, 49
why long is long, 23
win only by ceding control, 106
win or take, 44
winner, 3
winning loony move, 94
Winning Ways, xi, xii, 45, 67, 125, 126
W.K., WhizKid, 78
Wolfe, David, 125

X

XOR = exclusive-or, 49, 50